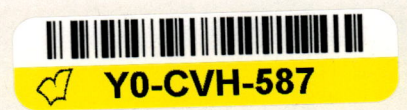

DISCARDED

THE USE OF GREASE AS AN ENGINEERING COMPONENT

© *The Institution of Mechanical Engineers 1970*

The Institution of Mechanical Engineers

Proceedings 1969–70 · Volume 184 · Part 3F

THE USE OF GREASE AS AN ENGINEERING COMPONENT

A Symposium arranged by the
Tribology Group
of the Institution of Mechanical Engineers
19–20th February 1970

1 BIRDCAGE WALK · WESTMINSTER · LONDON S.W.1

CONTENTS

		PAGE
Introduction		vii
Paper 1	Film thicknesses in elastohydrodynamic lubrication of rollers by greases, by A. Dyson, M.A., and A. R. Wilson, B.Sc., C.Eng., M.I.Mech.E.	1
Paper 2	Calculation of the effect of the compressibility of grease on the performance of a twin-line dispensing system, by J. F. Hutton, B.Sc.	12
Paper 3	Some designs for mountings incorporating two rolling bearings and a grease relief system, by G. W. Mullett, Wh.Sc., C.Eng., M.I.Mech.E.	17
Paper 4	IP dynamic anti-rust test, lubricating greases, by F. E. H. Spicer, C.Eng., M.I.Mech.E.	23
Paper 5	Apparent (dynamic) viscosity and yield strength of greases after prolonged shearing at high shear rates, by Ir G. J. Scholten	32
Paper 6	Greases as lubricants for metal and plastic-lined plain bearings, by J. G. M. Sallis, B.Sc., and W. H. Wilson, B.Sc.(Eng.), C.Eng., M.I.Mech.E.	40
Paper 7	A survey of rolling-bearing failures, by A. W. Morgan, B.Sc., and D. Wyllie, B.Sc., Ph.D.	48
Paper 8	Rheological behaviour of a new high-temperature synthetic grease, by A. E. Yousif, B.Sc., M.Sc., and K. D. Bogie, M.Sc., Ph.D. (*Graduate*)	57
Paper 9	Some failures of grease-lubricated rolling-element bearings, by E. D. Yardley, B.Sc., M.Sc., Ph.D., and W. J. J. Crump, B.Sc.	63
Paper 10	Some problems in the development of a high-performance grease for industrial rolling bearings, by H. D. Moore, B.Sc., J. W. Pearson, B.A., and N. A. Scarlett	74
Review Paper	Grease lubrication: a review of recent British papers, by P. L. Langborne, C.Eng., M.I.Mech.E.	82
Review Paper	Review of recent U.S.A. publications on lubricating grease, by R. S. Barnett	87
Discussion and communications		94
Authors' replies		102
Summary of discussion on report papers		107
List of delegates		108
Index to authors and participants		109
Subject index		110

The Use of Grease as an Engineering Component

A SYMPOSIUM on The Use of Grease as an Engineering Component, arranged by the Tribology Group of the Institution of Mechanical Engineers, was held at 1 Birdcage Walk, Westminster, London, S.W.1, on 19–20th February 1970. 118 delegates registered to attend.

The Symposium was opened by Mr M. J. Neale, Wh.Sch., B.Sc.(Eng.), C.Eng., F.I.Mech.E. Ten papers, two review papers, and a number of report papers were presented, as follows:

Session 1, Part 1. Chairman: Mr M. J. Neale. Review papers.
Session 1, Part 2. Chairman: Mr R. P. Langston. Papers 3, 2, and 6.
Session 2. Chairman: Mr R. P. Langston. Papers 7 and 9.
Session 3. Chairman: Mr T. I. Fowle, B.Sc.(Eng.), C.Eng., M.I.Mech.E. Papers 8, 5, 4, 1, and 10.
Session 4. Chairman: Mr J. A. Robertson, B.Sc., C.Eng., F.I.Mech.E. Report papers

The report papers, which are not included in the Proceedings, are listed below. A summary of the discussion on them is printed on p. 107.

Grease problems in aircraft bearings, by R. Cosher.
The distribution and application of lubricating greases as an engineering function, by D. R. Parkinson
Silicone greases, by K. MacKenzie
Grease-lubricated gear units, by I. S. Roberts
Grease lubrication for large industrial electric motors, by J. K. Vose
Grease pumpability, by A. G. Jackson
The lubrication of office equipment, by M. Marini
Grease for high-temperature use, by P. O. Trevalion
Grease lubrication in rolling bearings, by D. G. Hjertzen and R. A. Clarke
A relubrication system for rolling bearings, by C. W. Foot
The application of grease to the girth-wheel drive of rotary kilns by spraying, by P. C. Day

The members of the Organizing Committee were: Mr Robertson (Chairman), Mr Fowle, Mr Langston, Mr W. C. Pike, M.Sc., C.Eng., M.I.Mech.E., and Mr D. Scott, C.Eng., M.I.Mech.E.

Paper 1

FILM THICKNESSES IN ELASTOHYDRODYNAMIC LUBRICATION OF ROLLERS BY GREASES

A. Dyson*   A. R. Wilson*

The object of this work was to measure the thickness of the films formed by greases under conditions of elastohydrodynamic lubrication. The thicknesses of the films were estimated from measurements of the electrical capacity between two discs. In general, the mineral oils from which the greases were made formed films whose thickness did not vary with time. The films formed by the greases were initially thicker than those formed by the corresponding base oils but, after continuous running, became thinner than the oil films. Evidence is offered that the differences observed between oils and greases are related to the viscoelastic properties of the lubricants.

INTRODUCTION

THE FORMATION of the film of lubricant necessary for the correct functioning of heavily loaded components such as gears, ball and roller bearings, etc., depends on a mechanism known as elastohydrodynamic lubrication. This form of lubrication is at its simplest in a disc machine, and the thickness of the film of lubricant formed in such a machine may be estimated from measurements of the electrical capacity between the two discs.

This method has been used at the authors' laboratory (1)–(3)‡ but the lubricant has always been liquid. However, since most of the rolling contact bearings in service are lubricated by grease, we decided to investigate the thickness of the films formed between the discs when they were lubricated by greases instead of oils or other liquids. In this paper the thicknesses of the films formed by a number of greases are compared with those formed by the oils from which the greases are made.

Although the geometry of the contact between the discs in the disc machine differs in many respects from that of the contact between the various elements of a ball or roller bearing, it was nevertheless thought that the results would be of value as an indication of the thickness of films which would be generated in a practical bearing, and of the effects of factors such as oil viscosity, soap content, rolling speed, load, running time and shut-down. Although the greases used were specially made for the investigation, for reasons explained later, and were somewhat simpler in their formulation than normal commercial greases, there is no reason to suppose that their mechanical properties differed essentially from those of greases formulated along more conventional lines.

EXPERIMENTAL

Disc machine

The disc machine shown in Fig. 1.1 has been described elsewhere (1), and is essentially similar to that used by Crook (4). The discs were of case-hardened En 34 steel, 76·2 mm (3 in) in diameter and 25·4 mm (1 in) in width. To accommodate small relative displacements in the axial direction, the edges of one disc were rounded to give an effective contact width of 22·2 mm (0·875 in). The discs were ground to an eccentricity of less than 2·5 μm (0·0001 in) and to a surface roughness of 0·0375–0·05 μm (1·5–2 μin) c.l.a. The finish was improved to better than 0·02 μm (0·8 μin) by polishing with diamond paste. The discs were always in pure rolling contact.

The discs were electrically insulated from each other for purposes of capacitance measurement and an iron–constantan thermocouple was embedded 3·2 mm (0·125 in) below the surface of each disc. The thermocouples were connected to a moving-chart recorder through slip-rings. During the experiments the temperature recorded on the chart was maintained at 60°C ± 2 degC by adjustment of the position of an infra-red heating lamp or of a jet of cold compressed air directed at the discs. The discs were loaded together by means of a dead weight

---

The MS. of this paper was received at the Institution on 6th May 1969 and accepted for publication on 17th July 1969. 22
* Shell Research Ltd, Thornton Research Centre, P.O. Box 1, Chester, CH1 3SH.
‡ References are given in Appendix 1.2.

Fig. 1.1. Disc machine

acting through a split bell-crank lever and push-rod system. The electrical capacity between the discs was measured by a radio-frequency bridge supplied from a 19-kHz oscillator, and connections between the bridge and the discs were made through further slip-rings.

The lubricants used in this work included greases and the mineral oils from which these greases were made. A description of the composition and properties of the lubricants is given later.

Estimates of film thickness

For the interpretation of the capacity measurements in terms of film thicknesses, it was assumed that the discs were of the same shape as in the Hertzian case of dry contact, with the addition of a constant separation, h_0. Initially, it was assumed that the inlet and Hertzian sections were full of lubricant and that the film divided at the outlet edge of the Hertzian contact zone into two films, each of thickness $h_0/2$, and each adhering to one of the discs. The remainder of the outlet film was taken to be air. This condition will be referred to as the 'full inlet' condition, and it has been assumed throughout in the presentation of the results. Further information on the method of interpreting the experimental observations is given in reference (1).

During the course of the work it became necessary to admit the possibility that the inlet film would not always be completely full. The extreme case of an empty inlet film would be when its configuration was the same as that previously assumed for the outlet film, and this is referred to as the 'empty inlet' condition. In practice the film thickness, h_0, would be expected to lie between the two extreme values calculated for the 'full inlet' and 'empty inlet' conditions. An example of the effect on the estimated value of film thickness of the assumption made regarding the filling of the inlet section is shown in Fig. 1.2.

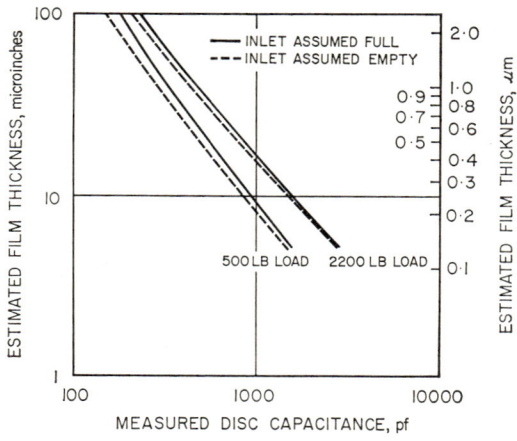

Fig. 1.2. Influence of the filling of the inlet section on the relation between film thickness and capacitance for grease G_2

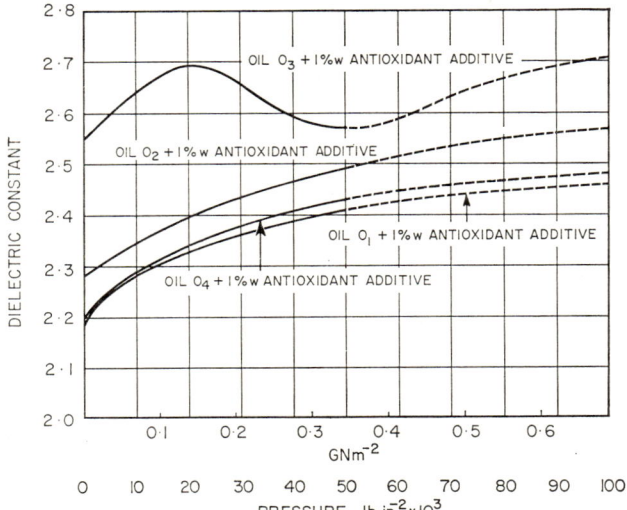

Fig. 1.3. Effect of pressure on the dielectric constant of test oils at 60°C and 19 kHz

Measurements of dielectric constants

The interpretation of the capacitance measurements in terms of film thickness depends not only on the degree of filling by the lubricant as discussed above, but also on the dielectric constant of the lubricant at the temperature and pressures existing in the disc machine. The dielectric constants of the component oils at atmospheric pressure were measured in a modified Henley dielectric cell, while measurements at pressures up to 345 MN/m² (50 000 lbf/in²) were made with a special cell, as described in reference (5). For the interpretation of capacitance measurements made at the higher of the two loads used, the dielectric constant was required at the mean Hertzian pressure of approximately 689 MN/m² (100 000 lbf/in²), and this was estimated by graphical extrapolation from the results covering the range of pressure from atmospheric to 345 MN/m² (50 000 lbf/in²). The extrapolation is shown in Fig. 1.3.

The extrapolation relating to oil O_3 in Fig. 1.3 must be regarded as speculative, but the mechanism by which the dielectric constants of viscous polar liquids are thought to be affected by pressure (5) demands an increase in dielectric constant over the range of extrapolation. The assumed value of dielectric constant at 689 MN/m² (100 000 lbf/in²) may be in error by a few per cent, and this may introduce a corresponding, though rather smaller, error into the conversion of capacitance measurements to values of film thickness for this oil. It will not, however, affect determination of the ratio of the thickness of a film formed by grease made from this oil to that formed by the oil alone, since the error in dielectric constant is common to both lubricants.

The dielectric constants of the greases at atmospheric pressure were measured in a simple parallel-plate condenser. In this condenser a rectangular upper electrode, 101·6 mm (4 in) long and 50·8 mm (2 in) wide, was separated from a lower electrode of much greater area by three Tufnol pins of 6·35 mm (0·25 in) diameter. These pins, arranged at the points of an isosceles triangle, protruded 1·02 mm (0·040 in) from the face of the lower electrode. Before use the condenser was calibrated with fluids of known dielectric constant.

No attempt was made to measure the dielectric constant of the greases at high pressures owing to experimental difficulties and to an uncertainty in the application of the results that will now be discussed.

Composition and properties of the greases

Practical greases, as usually formulated, contain many polar additives and their dielectric constants are high, with measured values ranging up to 5. Furthermore, these dielectric constants are known to decrease with increasing rate of shear (6)–(8), and this has been attributed to the progressive breakdown with shear of the three-dimensional fibrous structure of the soap particles in the greases. This variation of dielectric constant with shear rate introduces an uncertainty in the estimates of film thickness, and it was desired to reduce this uncertainty to a minimum.

Table 1.1. Properties of the base oils

Oil	O_1	O_2	O_3	O_4
Specific gravity, 60°F/60°F	0·886	0·909	0·955	0·873
Kinematic viscosity, cS, 100°F	32·6	138·6	578·0	27·5
210°F	4·66	10·2	19·9	4·73
Kinematic viscosity index	40·0	39·0	−10·0	98·0
Dynamic viscosity, cP, 60°C	11·44	34·05	108·0	10·57
Dielectric constant at atmospheric pressure, 20°C, 19 kHz	2·26	2·34	2·65	2·27
60°C, 19 kHz	2·20	2·28	2·55	2·21

Note: 1%wt of an anti-oxidant additive was added to each of the oils before the properties were measured.

Table 1.2. Composition and properties of the greases

Code	G_1	G_2	G_3	G_4
Composition of mix (1b)				
HCOFA	6·00	6·00	3·60	5·25
LHM	0·88	0·88	0·53	0·77
Anti-oxidant additive	0·57	0·57	0·60	0·50
O_1	49·55	—	—	—
O_2	—	49·55	55·27	—
O_3	—	—	—	43·48
Total	57·00	57·00	60·00	50·00
Volume concentration of soap in finished grease, per cent	9·8	10·0	5·6	10·4
Dielectric constant, atmospheric pressure, 20°C, 19 kHz	2·64	2·69	2·57	2·93
Worked penetration number*	196	193	253	192

* IP 50/64: This differs only in small details from ASTM method D217.

With this in mind the greases used in this investigation were made in the simplest possible way. Commercial fatty acid, derived from hydrogenated castor oil (HCOFA, consisting of approximately 90%wt hydroxystearic acid and 10%wt stearic acid, with a nominal equivalent weight of 300 based on a saponification value of 187 mg KOH/g approximately) and the appropriate amount of lithium hydroxide monohydrate (LHM) were heated to 140°C in the selected base oil in an autoclave. When the pressure was released, the water of hydration and that produced by the reaction were driven off; the soap dissolved in the oil after heating to 200°C. The rates of subsequent cooling and stirring were adjusted to give the structure and penetration required. The properties of the base oils are given in Table 1.1, and the compositions of the mixes and the properties of the finished greases in Table 1.2.

To prevent oxidation of the oil during the manufacture of the grease, 1·0%wt of an anti-oxidant additive was

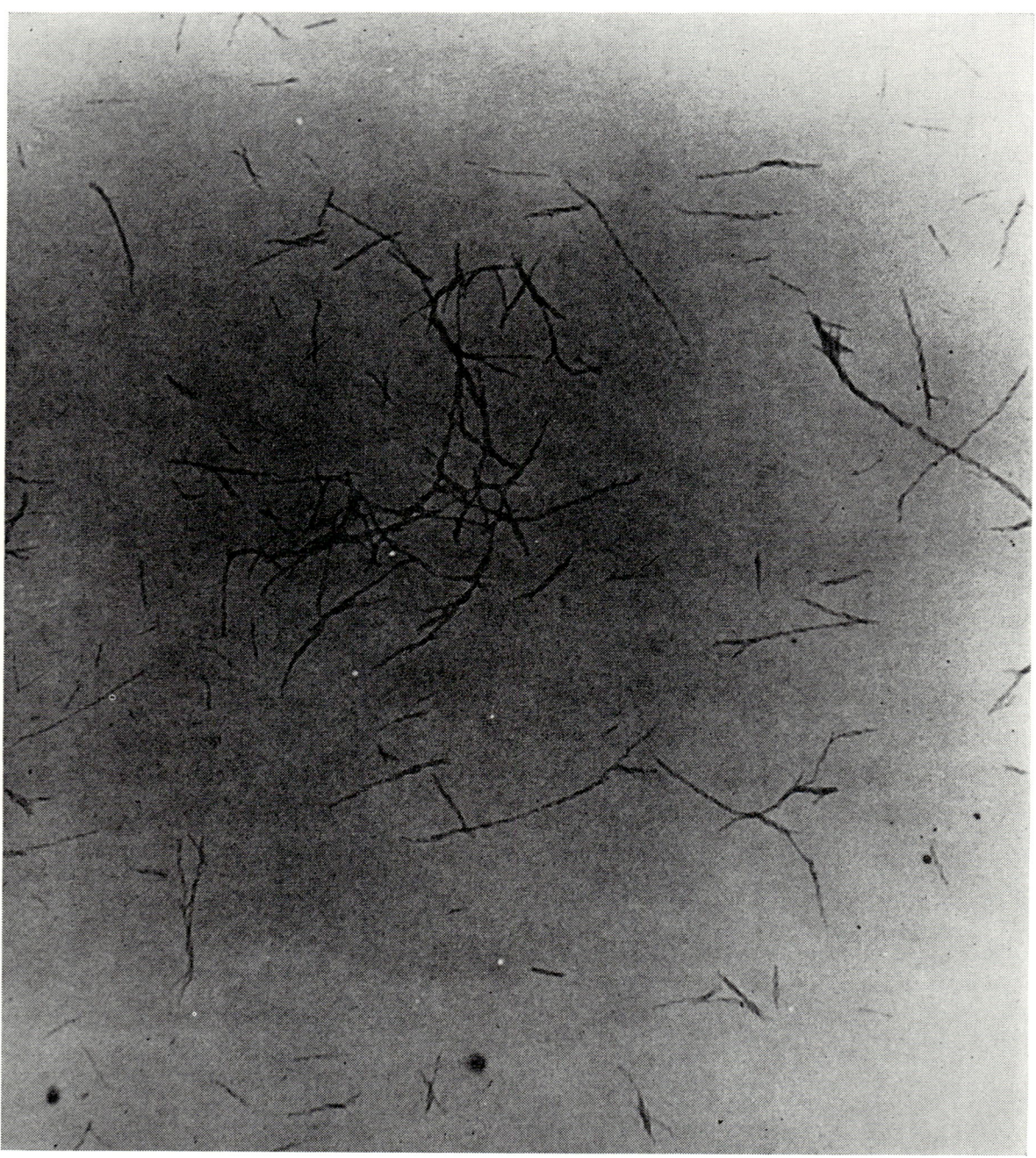

Fig. 1.4. Original fibrous structure of the soap in grease G_2. ×25 000

incorporated in each mix. The plan was that a comparison between the behaviour of greases G_2 and G_3 would show the effect of a change in the soap content with a constant base oil viscosity, while a comparison of G_1, G_2, and G_4 would show the effect of base oil viscosity at a constant soap content. The calculation of the volume fraction of soap in the finished grease is explained in Appendix 1.1.

The uncertainty in the dielectric constant must now be discussed in greater detail. The original fibrous structure of the soap in the grease, depicted in Fig. 1.4, will be broken down under the conditions of high shear stress encountered in the inlet zone of the disc machine, and the extreme end condition would be a dispersion of uniform spherical particles of soap in the oil medium. Fig. 1.5, which shows the condition of soap particles on one of the discs after many revolutions through the loaded contact, suggests that this end condition is probably

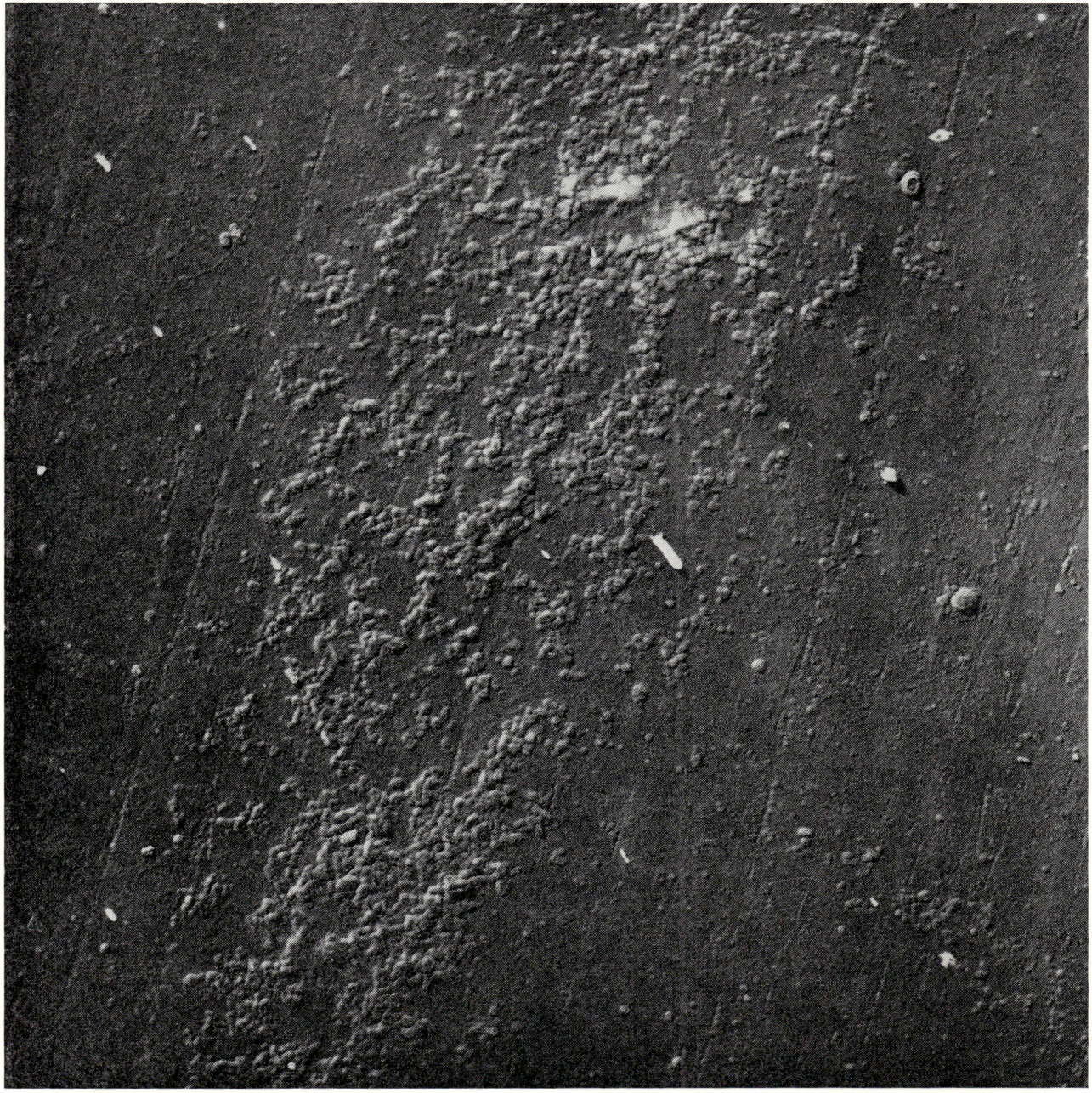

Fig. 1.5. Carbon replica of soap particles on surface of a disc after 15 min running in the disc machine. ×25 000

attained. The dielectric constant, ϵ_3, of such a suspension would be (9)

$$\epsilon_3 = \epsilon_1 + \Delta\epsilon_1$$

where

$$\Delta\epsilon_1 = \frac{3c\epsilon_1(\epsilon_2 - \epsilon_1)}{\epsilon_2 + 2\epsilon_1}$$

and ϵ_1 is the dielectric constant of the oil medium, ϵ_2 is the dielectric constant of the soap, and c is the volume fraction of soap in the suspension.

This equation is correct to the first order in c, so that it should be applicable to the greases used in the present work where the highest volume fraction used was approximately 0.1 (10% vol.).

The dielectric constant, ϵ_2, of the soap was measured with the same parallel-plate capacitor as was used for the greases, and the value found was 3·3. The calculated values of the dielectric constants, ϵ_3, of the suspensions are compared in Table 1.3 with the experimental values, ϵ_4, for the greases, measured at zero rate of shear. The effective values for the greases when subjected to very high rates of shear in the disc machine will be expected to lie between these two limits. The range of uncertainty is small, of the order of 10 per cent, and the effective dielectric constants of the greases in the disc machine at 20°C at atmospheric pressure are arbitrarily taken to have the values of ϵ_3. The uncertainty was much greater for some fully formulated greases that were tested in some preliminary work.

Experimental measurements on greases at high pressures would be subject to a similar uncertainty regarding the effect of shear rate. Consequently, effective values are calculated on the assumption that the contribution, $\Delta\epsilon_1$, made by the addition of soap to the oil medium, is constant with pressure, the value of $\Delta\epsilon_1$ for each grease being added to the dielectric constant of the appropriate base oil at each pressure. The values obtained are shown in Table 1.4.

Finally, it is necessary to consider the expected effect of the thickening of the oil by the suspension of soap. It has been argued that the fibrous structure of the soap will be broken up by the shear field, and the result in the extreme case would be a suspension of spherical soap

Table 1.3. Dielectric constants of greases at 20°C at atmospheric pressure

Code	G_1	G_2	G_3	G_4
Dielectric constant, ϵ_1, of base oil and anti-oxidant	2·26	2·34	2·34	2·65
Volume fraction of soap, c	0·098	0·100	0·056	0·104
Calculated increase in ϵ_1 due to presence of spherical soap particles, $\Delta\epsilon_1$	0·09	0·09	0·05	0·06
Calculated dielectric constant, ϵ_3, of suspension of spherical soap particles	2·35	2·43	2·39	2·71
Measured dielectric constant, ϵ_4, of grease at rest	2·64	2·69	2·57	2·93
Ratio ϵ_4/ϵ_3	1·12	1·11	1·08	1·08

Table 1.4. Dielectric constants of lubricants at 60°C at various pressures

Lubricant	Dielectric constant at Atmospheric pressure*	345 MN/m² (50 000 lbf/in²)	689 MN/m² (100 000 lbf/in²)‡
O_1	2·19	2·41*	2·46
O_2	2·28	2·50*	2·57
O_3	2·55	2·57*	2·70
O_4	2·20	2·43*	2·48
G_1	2·28	2·50†	2·55
G_2	2·37	2·59†	2·66
G_3	2·33	2·55†	2·62
G_4	2·62	2·65†	2·77

* By direct measurement.
† By calculation, as explained in the text.
‡ By extrapolation.
Note: 1% wt of anti-oxidant additive was present in each lubricant.

particles in the oil medium. The maximum thickening effect would be apparent if these soap particles were rigid; this is certainly an over-estimate, but probably the best available. The thickening power of such a suspension of rigid spherical particles is given to the first order in the volume concentration, c, by the Einstein relation

$$\eta_3/\eta_1 = 1 + 2·5c$$

where η_3 is the viscosity of the suspension, and η_1 is the viscosity of the oil medium.

An expression which is accurate to a higher order in c is (10)

$$\eta_3/\eta_1 = (1 - 2·5c)^{-1}$$

Since the film thickness in elastohydrodynamic lubrication is proportional to the viscosity raised to the power of 0·7 (11), the maximum expected increase in the film thickness, due to the thickening effect of the broken-down soap particles, is given by a factor

$$(1 - 2·5c)^{-0·7}$$

This increase in film thickness is 22 per cent for greases G_1, G_2, and G_4 with approximately 10% vol. concentration of soap, and 11 per cent for G_3 with 5·6% vol. soap.

Experimental procedure

At the start of a run the discs were rotated slowly under no load and cleaned by tissues soaked with a light paraffinic solvent. They were then brought up to test speed and the lubricant was applied.

The greases were supplied to the inlet to the discs as lumps on the end of a plastic spatula. The lumps were approximately 0·5 ml in volume, and although no systematic measurements were made it was observed that the size of the lump had no appreciable effect on the capacitance measurements. The oils were applied in drops from the end of a glass rod, the quantity of oil (approximately 0·05 ml) being considerably less than the quantity of grease normally applied.

After the application of the grease or oil the load was applied and the position of the infra-red heating lamp or of the jet of cooling air was then adjusted to give the required temperature of 60°C, as indicated by the thermocouples embedded in the discs.

A fresh drop or lump of lubricant was then applied and the run was timed from this moment. This procedure was justified by the observation that, although the measured capacity varied with time, the application of fresh lubricant always brought the capacity back to the starting value.

The behaviour of the lubricants was examined under three different operating procedures:

(1) Speed and load were maintained constant during any one test. Different tests were run at speeds of 1600 and 400 rev/min and at loads of 9·79 and 2·45 kN (2200 and 550 lbf). These loads gave mean Hertz stresses of 714 and 357 MN/m^2 (103 600 and 51 800 lbf/in^2) respectively.

(2) Load was maintained constant at 9·79 kN (2200 lbf) but the disc machine was run at 1600 and 400 rev/min during alternate periods of 15 min.

(3) A constant load of 9·79 kN (2200 lbf) was applied at a speed of 1600 rev/min but the disc machine was stopped and immediately restarted under load at intervals of 15 min.

The temperature of the discs was maintained at 60°C± 2 degC in all three procedures.

RESULTS

The greases and oils listed in Table 1.4 were examined in accordance with the three test procedures. Not all of the lubricants, however, were examined under all conditions. The film thicknesses deduced from capacitance measurements are shown in Figs 1.6–1.16.

Results obtained in continuous tests at constant speed and load (Procedure 1) are shown in Figs 1.6–1.10 and in each figure the film thicknesses of one of the greases, plotted as a function of time, are compared with those given by the oil from which the grease was made. Figs 1.6–1.9 relate to greases G_1, G_2, G_3, and G_4 respectively, while Fig. 1.10 shows results obtained with grease G_2 at the lower load. In all cases the films formed by the greases were initially thicker than those of their component oils by an amount that appears to be consistent with the

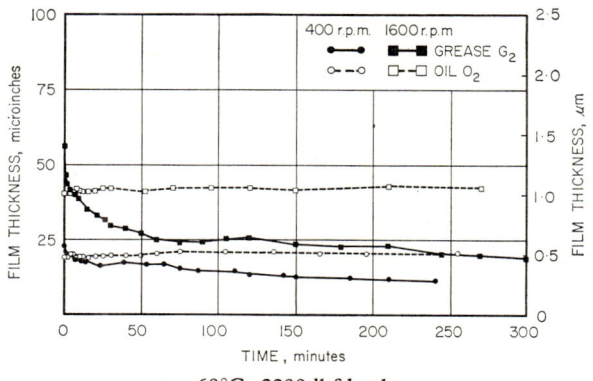

60°C; 2200 lbf load.

Fig. 1.7. Variation with time of film thickness of grease G_2 and base oil O_2

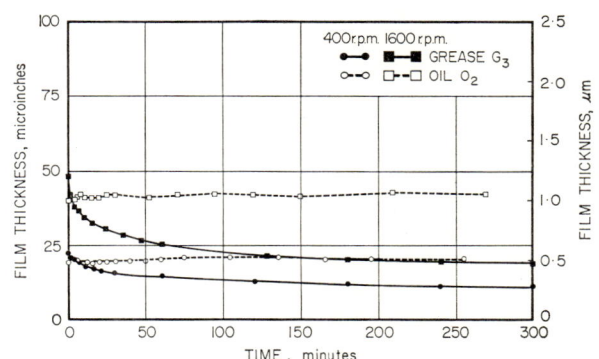

60°C; 2200 lbf load.

Fig. 1.8. Variation with time of film thickness of grease G_3 and base oil O_2

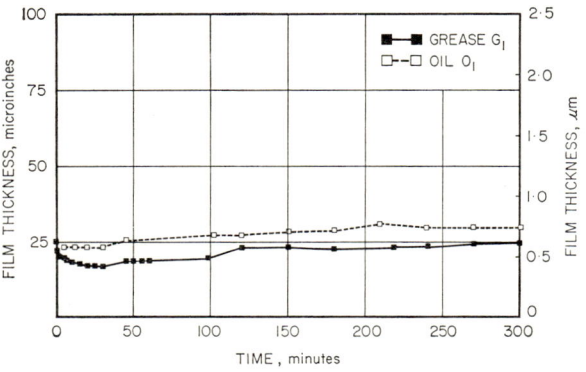

60°C; 2200 lbf load; 1600 rev/min.

Fig. 1.6. Variation with time of film thickness of grease G_1 and base oil O_1

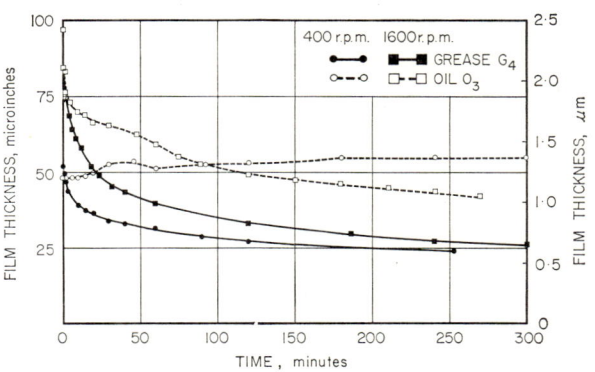

60°C; 2200 lbf load.

Fig. 1.9. Variation with time of film thickness of grease G_4 and base oil O_3

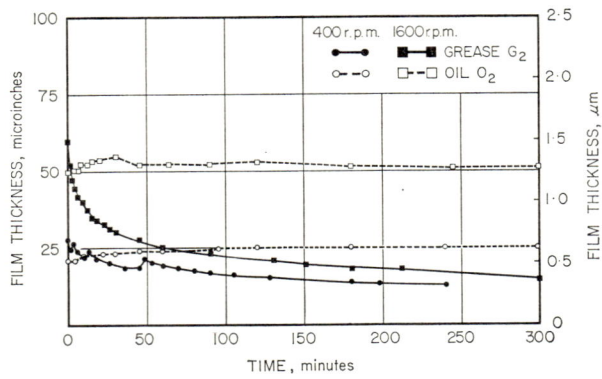

60°C; 550 lbf load.

Fig. 1.10. Variation with time of film thickness of grease G₂ and base oil O₂

thickening power expected for a uniform suspension of rigid soap particles. The greater the viscosity of the base oil, the rotational speed, and the soap content of the grease, the thicker the initial film. The difference in penetration numbers of greases G₂ and G₃ was not reflected by any marked difference in behaviour. A comparison of Fig. 1.7 with Fig. 1.10 shows that, as would be expected on theoretical grounds, the effect of load on film thickness was not large.

Whereas the thickness of the oil films was, in general, approximately constant throughout a test, the grease films generally became thinner with time at both speeds and loads and the thickness rapidly became less than that of the corresponding oil film. With one exception the grease films seemed to reach equilibrium at a thickness approximately 40 per cent of the initial value. At any stage during an experiment a further application of grease would bring the capacity down to the initial value, i.e. the film thickness up to the initial value.

An exception to the general pattern of behaviour of the greases is shown in Fig. 1.6 by grease G₁. The initial rate of decrease in film thickness was smaller than that observed with the other greases at the same rotational speed and the film thickness subsequently increased to approximately the initial value. The significance of this increase, which was also observed with the oil on which the grease was based, is discussed later in this paper.

As shown in Figs 1.7–1.10, the thickness of the films formed by the oils generally remained approximately constant with an apparent tendency to increase slightly with time. This tendency, however, probably reflects a reduction in capacity caused by a progressive partial emptying of the inlet section of the Hertzian contact rather than a real change in thickness. At the start of a test the inlet must be full, but it was observed that the film thickness calculated at the end of a test on the 'empty inlet' hypothesis was approximately equal to that calculated at the start of the test on the 'full inlet' assumption.

There were two exceptions to the general observation that oils maintain films of approximately constant thickness. First, Fig. 1.9 shows that the most viscous oil, O₃, at the higher speed formed a film which decreased in thickness with time to half the initial value. This may be related to the effect of centrifugal force in throwing off the oil against the surface tension. It occurs when the initial thickness, $h_0/2$, of the film on the circumference of each disc is of the same order as the theoretical maximum equilibrium film thickness, h_1, that can be retained by the surface tension against centrifugal force. This thickness is given by the relation

$$h_1 \sim T/\rho u^2$$

where T is the surface tension, ρ the density of the lubricant, and u is the peripheral speed of the disc.

The second exception is seen in Fig. 1.6 with an oil of low viscosity, O₁. With this oil the decrease in capacity with time was five times greater than that attributable to emptying of the inlet section, and an explanation must be sought for an apparently real increase in film thickness. A tentative suggestion is that the most volatile constituents of a light oil may evaporate from a film 0·25 μm (10 μin) thick on the surface of a disc at 60°C rotating in air at a peripheral speed of 6·4 m/s (21 ft/s). Such evaporation would lead to a gradual increase in viscosity of the reserve of oil in the inlet to the Hertzian contact and hence to an increase in film thickness. Conversely, an oil from which the most volatile constituents had been removed previously would not be expected to behave in the same way. This was confirmed by a test run on oil O₄. This is a mineral oil from which the light ends have been stripped and it is intended for use in high-vacuum systems. Its vapour pressure at 60°C is $2·1 \times 10^{-4}$ torr compared with $2·3 \times 10^{-3}$ torr for O₁, although the viscosities of the two oils are similar. Fig. 1.11 shows that oil O₄ maintained a film of constant thickness, and this result supports the suggestion that the increase of film thickness with time, experienced with O₁, was due to selective evaporation of the lighter fractions.

Results obtained under procedure 2, i.e. with speed cycled between 1600 and 400 rev/min, are shown for

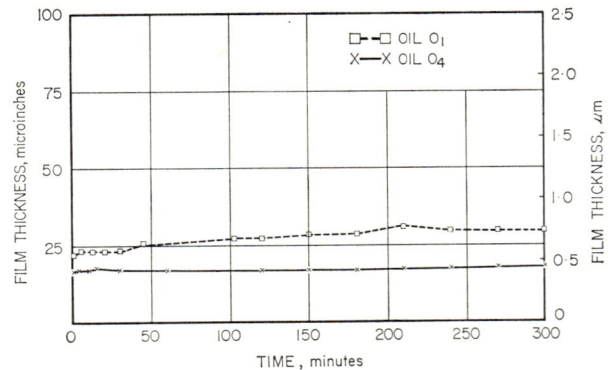

60°C; 2200 lbf load; 1600 rev/min.

Fig. 1.11. Variation with time of film thickness of oils O₁ and O₄

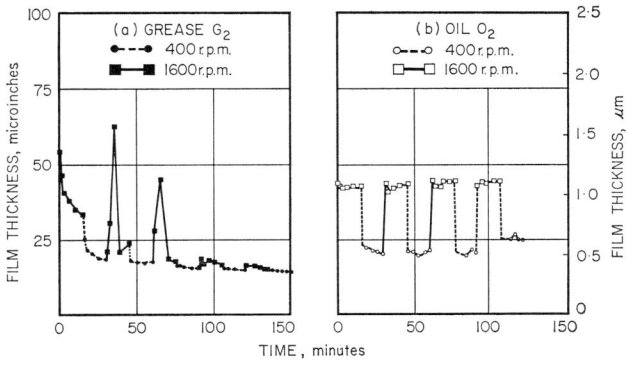

Fig. 1.12. Effect of variation of speed on film thickness of (*a*) grease G$_2$ and (*b*) base oil O$_2$

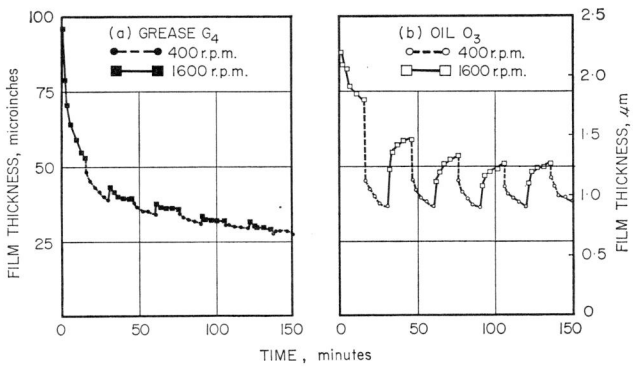

Fig. 1.13. Effect of variation of speed on film thickness of (*a*) grease G$_4$ and (*b*) base oil O$_3$

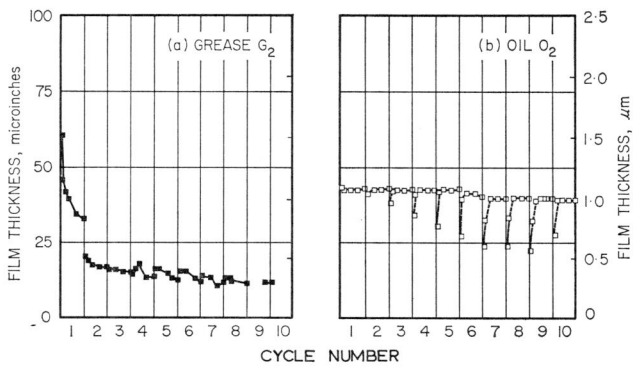

Fig. 1.14. Effect of repeated shut-down under load on film thickness of (*a*) grease G$_2$ and (*b*) base oil O$_2$

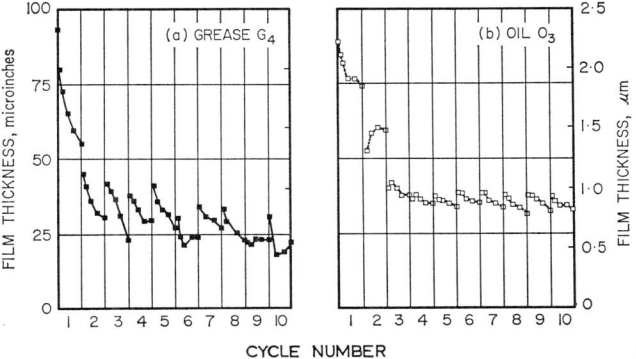

Fig. 1.15. Effect of repeated shut-down under load on film thickness of (*a*) grease G$_4$ and (*b*) base oil O$_3$

greases G$_2$ and G$_4$ in Figs 1.12 and 1.13. The thickness of the oil films always follows the change in speed with a delay that increases with increasing oil viscosity. The effect of changes of speed on the thickness of the grease films gradually decays with time, the final thickness tending towards an equilibrium value which is approximately equal to the value that would have been reached if the machine had been operated continuously at the lower speed. At the start of the second and third periods at the higher speed in the run on grease G$_2$ (Fig. 1.12), the film thickness recovered temporarily to almost its initial value at the beginning of the first period, but rapidly declined towards the end of the periods. The explanation for this occurrence is not known.

In this work it was noted that the effect of stopping and restarting the disc machine in the unloaded condition was very small both for oils and greases. However, the effect of stopping and restarting under load, as in procedure 3, was more marked and is shown for greases G$_2$ and G$_3$ in Figs 1.14 and 1.15. The general effect of repeatedly stopping and restarting under load was to accelerate the reduction in film thickness with time that would occur in continuous running. In one case, with grease G$_2$, the decrease in film thickness caused an increase in frequency of electrical contact between the discs such that measurements of capacitance became impracticable after the ninth shut-down. An exception to the general pattern of behaviour is observed in Fig. 1.14 with oil O$_2$, which recovered its initial value of film thickness after a delay of up to 5 min after a restart.

DISCUSSION

Many bearings run for years on the same charge of grease, but the maximum duration of the tests reported here was 5 hours. The results, therefore, refer only to a very small fraction of the service life. Furthermore, the quantity of grease applied to the discs was much smaller than that contained in a bearing. Within these limitations it is believed that the work gives some evidence about the nature and behaviour of the lubricant in the grease-lubricated bearing.

The fact that grease films are initially thicker than the corresponding oil films, and the very different time behaviour, show that the lubricating agent is something other than the oil alone. The fair agreement between the observed and expected thickening effect of the soap suggests that the lubricating medium consists of a suspension of soap in the oil, the initial fibrous structure of the soap having been broken down by shear into a suspension of particles which are approximately spherical in shape. Fig. 1.5 shows an electron micrograph, at a magnification of ×25 000, of a carbon replica of soap particles on the surface of a disc after 15 min running in the disc machine. The soap in the original grease was arranged in a fibrous structure, as shown in Fig. 1.4, but this has been almost completely destroyed by working in the machine.

The initial thinning of the grease films with continued running may be due to the progressive breakdown with shear of the structure of the grease, but this cannot be the complete explanation since the films formed by greases eventually become thinner than those formed by the corresponding oils. An explanation of this observation is offered along the following lines.

The maximum shear rate in the inlet zone gradually increases as the lubricant approaches the conjunction. The structure of the grease will be progressively broken down, but there will be a region in which, although the shear rate is quite high, the initial fibrous structure of the grease is preserved to some extent. The grease in this region will behave as a viscoelastic material and will exhibit normal stress differences in shear.

With axes parallel to the direction of motion and to the velocity gradient, the stress system in a Newtonian liquid subject to a simple shear consists of a shear stress and an isotropic normal stress, i.e. the hydrostatic pressure. This is not true for a sheared viscoelastic liquid; the normal stress component is no longer isotropic, and the principal effect is equivalent to a tension along the stream lines.

It is suggested that these normal stress differences cause a migration from regions of high shear rate to regions of low shear rate, which may be related to the 'clearing' or 'channelling' effect observed in rolling contact bearings. In the disc machine this migration would cause a partial emptying of the inlet section by leakage towards the sides of the discs, and a subsequent decrease in film thickness.

Evidence in support of the importance of normal stress differences was obtained by testing a solution of 5%wt polyisobutene (Oppanol B50 weight, average molecular weight 365 000) in oil O_2 with 1%wt anti-oxidant additive. Such a solution is viscoelastic and would show normal stress differences in shear. The thickness of the film formed in the disc machine by this lubricant decreased with time in a manner similar to that observed for greases, as shown in Fig. 1.16. The observation that the films rapidly become thinner than those given by the base oil alone again shows that shear breakdown of the polymer cannot be a complete explanation.

ACKNOWLEDGEMENTS

The authors wish to thank Mr H. D. Moore for his expert guidance on the characteristics of greases, Messrs K. Greenway and G. W. Turner for preparing the greases and for a lot of unreported but essential preliminary work, Messrs R. B. Bird and G. Rooney for the preparation of the electron micrographs, and, particularly, Mr W. J. Cairney for making most of the experimental measurements.

APPENDIX 1.1

CALCULATION OF THE VOLUME FRACTION OF SOAP IN THE FINISHED GREASES

The soap is formed by reaction of HCOFA with LHM:

$$RCOOH + LiOH \cdot H_2O \rightarrow RCOOLi + 2H_2O$$

The water is driven off during manufacture. The saponification number of 187 mg KOH/g for the HCOFA corresponds to an equivalent weight of 300 and the equivalent weight of the lithium soap is therefore 306. The equivalent weight of LHM is 42, and slight excess of base is used. Thus, in the manufacture of grease G_1 the 6·00 lb of HCOFA (Table 1.2) would react with

$$6 \cdot 00 \times \frac{42}{300} = 0 \cdot 84 \text{ lb LHM}$$

to produce

$$6 \cdot 00 \times \frac{306}{300} = 6 \cdot 12 \text{ lb lithium soap}$$

The weight of LHM remaining unchanged is

$$0 \cdot 88 - 0 \cdot 84 = 0 \cdot 04 \text{ lb}$$

The water from the excess LHM is lost during manufacture, and most of the excess lithium hydroxide is converted to lithium carbonate by carbon dioxide from the air, 0·04 lb LHM being equivalent to 0·04 lb lithium carbonate. The compositions by weight of the finished greases may now be calculated, and are given in Table 1.5.

The compositions by volume may now be calculated if the specific gravities are known. The specific gravity of the soap was determined as 1·0 by weighing in water, and that of lithium carbonate is given as 2·1 (*12*). The specific gravities of the oils were obtained from Table 1.1,

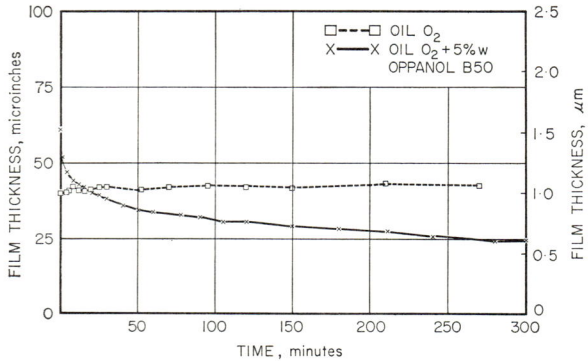

60°C; 2200 lbf load; 1600 rev/min.

Fig. 1.16. Effect on film thickness of the addition to the test oil of 5%wt Oppanol B50

Table 1.5. Calculated compositions of finished greases

Component	Weight, lb				Relative volume			
	G_1	G_2	G_3	G_4	G_1	G_2	G_3	G_4
Lithium soap	6·12	6·12	3·67	5·36	6·12	6·12	3·67	5·36
Lithium carbonate	0·04	0·04	0·03	0·03	0·02	0·02	0·01	0·02
Mineral oil + anti-oxidant	50·12	50·12	55·87	43·98	56·57	55·13	61·45	46·06
Total	56·28	56·28	59·57	49·37	62·69	61·27	65·13	51·44

the effect of differences in the proportion of anti-oxidant being ignored. The results are also given in Table 1.5, and the volume concentrations of soap in the finished greases, quoted in Table 1.2, are worked out from these results.

APPENDIX 1.2
REFERENCES

(1) DYSON, A., NAYLOR, H. and WILSON, A. R. 'The measurement of oil-film thickness in elastohydrodynamic contacts', *Symp. Elastohydrodynamic Lubrication, Proc. Instn mech. Engrs* 1965–66 **180** (Pt 3B), 119.

(2) DYSON, A. and WILSON, A. R. 'Film thicknesses in elastohydrodynamic lubrication by silicone fluids', *Lubrication and Wear Fourth Conv., Proc. Instn mech. Engrs* 1965–66 **180** (Pt 3K), 97.

(3) DYSON, A. and WILSON, A. R. 'Film thicknesses in elastohydrodynamic lubrication at high slide/roll ratios', *Proc. Instn mech. Engrs* 1968–69 **183** (Pt 3P), 83.

(4) CROOK, A. W. 'The lubrication of rollers—I', *Phil. Trans. R. Soc.* 1958 **A250**, 387.

(5) GALVIN, G. D., NAYLOR, H. and WILSON, A. R. 'The effects of pressure and temperature on some properties of fluids of importance in elastohydrodynamic lubrication', *Lubrication and Wear Second Conv., Proc. Instn mech. Engrs* 1963–64 **178** (Pt 3N), 283.

(6) BONDI, A. and PENTHER, C. J. 'Some electrical properties of colloidal suspensions in oils', *J. Phys. Chem.* 1953 **57**, 72.

(7) VINOGRADOV, G. V. and DEINEGA, YU. F. 'Electrical properties of lubricating greases in relation to peculiarities of their structure', *J. Inst. Petrol.* 1966 **52**, 279.

(8) BRIGHT, G. S. 'The use of electrical measurements as an aid to the understanding of fundamental grease structure', *N.L.G.I. Spokesman* 1966 **30** (3), 84.

(9) LANDAU, L. D. and LIFSHITZ, E. M. *Electrodynamics of continuous media* 1960, 47 (Pergamon Press).

(10) HERMANS, J. J. *Flow properties of disperse systems* 1953, Ch. I (North-Holland, Amsterdam).

(11) DOWSON, D. and HIGGINSON, G. R. 'New roller bearing lubrication formula', *Engineering, Lond.* 1961 **192**, 158.

(12) LANGE, N. A. *Handbook of chemistry* 1961, 10th edition (McGraw-Hill).

Paper 2

CALCULATION OF THE EFFECT OF THE COMPRESSIBILITY OF GREASE ON THE PERFORMANCE OF A TWIN-LINE DISPENSING SYSTEM

J. F. Hutton*

The compressibility of a grease can affect in two ways the time required for the operation of a twin-line dispensing system. First, during the application of pump pressure to one of the two main pipes of the system, time is taken to compress grease into the pipe in addition to that taken to supply grease to the bearings along the pipe. Secondly, during the next cycle, when that pipe is opened to the atmosphere and the other one is being pressurized, time is required for the compressed grease to flow back into the reservoir and thus for the pressure in the pipe to fall to a value sufficiently low to allow the dispensing valves to operate.

Formulae are given for calculating these time factors, and for typical modern practice times of about 2 or 3 min are obtained. This explains the observation that some twin-line systems cannot be successfully operated on cycle times as short as 3 min. The formulae may be useful in the design of twin-line systems.

STATEMENT OF THE PROBLEM

A TWIN-LINE centralized grease lubrication system is based on a pair of parallel, large-diameter pipes which extend from the grease pump to the end of the plant to be lubricated. At intervals along the pipes, and bridging them, there are valve blocks which connect through smaller pipes to the bearings requiring lubrication. Pump pressure is applied to the main pipes alternately, so that when one is being pressurized the pressure in the other is relaxing. Each valve block contains a shuttle valve and a metering piston. The shuttle valve directs grease to one or the other end of the piston, which displaces a metered quantity of grease through the outlet to the bearings. The whole operation is powered by the difference in pressure between the pressurized pipe and the relaxing pipe. When the last valve along the pipe has been operated, the pump is automatically switched off. Sometime later, according to a pre-arranged schedule, the pump is switched on again, but this time pressure is applied to the pipe which previously was relaxing. As pressure is built up, the valve blocks operate in turn, the pistons moving in the reverse direction.

The MS. of this paper was received at the Institution on 18th July 1969 and accepted for publication on 13th August 1969. 33
* Shell Research Ltd, Thornton Research Centre, P.O. Box 1, Chester CH1 3SH.

A cycle is the time interval between successive switchings-on of the pump, during which the valves operate once.

In designing a lubrication system, it is the practice of one manufacturer (Farvalube Ltd, Hereford) to select a standard pump which normally develops 1200 lb/in² (8·27 MN/m²) in pumping grease at a rate of 10 oz/min (4·72 g/s). The pressure drop permitted along the required length of main is 600 lb/in² (4·14 MN/m²), thus providing 600 lb/in² at the end to operate the last valve. With the aid of charts a pipe diameter is selected to give this pressure drop. The charts in use by Farvalube are the result of experience and are based on the viscosity/rate of shear characteristic of a fictitious 'average' grease. From calculations we have made from data supplied to us, the 'average' grease is similar to grade 2 lithium greases at about 15°C. The properties of this 'average' grease will be referred to again later (1)†.

Normally the twin-line system is trouble-free, provided the grease has viscosity characteristics similar to those of the 'average' grease. Nevertheless, difficulties have arisen in the pumping of such greases when the cycle time has been reduced to 3 min from the normal range of 10–30 min. The problem is manifest as a 'snowballing' of

† *The reference is given in Appendix 2.1.*

the pressure at the final valve, where the pressure continues to rise instead of oscillating between the designed 600 lb/in² on the pressure cycle and zero (or a value determined by the yield stress) on the relaxing cycle. There is a high margin of performance of the pump, and the end pressure can rise considerably before the pumps cut out.

The probable cause of 'snowballing' may lie in the fact that grease is compressible. Were grease incompressible, the relaxation of pressure in the pipe would occur at the speed of sound when the pump end is opened to atmosphere. In fact, an appreciable amount of grease is compressed into the pipe, and this must flow back into the reservoir to enable the pressure to normalize. A valve operates under a pressure differential of 400 lb/in², and at the end of the pipe 600 lb/in² pressure is normally available. Therefore, if the system is to work normally, the end pressure in the relaxing pipe must fall to below 200 lb/in² when the valve is due to operate. If the end pressure has not fallen sufficiently low, the pump will cause the actuating pressure to exceed 600 lb/in². On the next cycle the situation will be a little worse and a still higher actuating pressure will be required, and so on. 'Snowballing' is aggravated by the modern requirements for dispensing systems to have longer lengths of pipe and to operate on shorter cycles.

This paper presents some calculations which show that effects of compressibility can limit the cycle time. The formulae obtained may assist in the design of centralized lubrication systems to avoid 'snowballing', but it would be necessary to take into account all the differences between the proposed design and the simplified arrangements considered in the paper.

THEORY
The excess volume compressed into a pipe

```
        P_p     x    dx       P_e
Pump   |―――――――――|―|――――――――|  Last valve
end    |         |  |        |  end
                 L
```

Consider the element dx at length x from the pump end of the pipe, whose total length is L. P_p and P_e are the pressures at the pump end and last valve end, respectively. The volume of the element of grease is

$$dV = \pi R^2 \, dx$$

where R is the internal radius of the pipe, and its uncompressed volume is dV_0.

Therefore, the excess volume of grease is

$$dV_0 - dV = (dV)\beta P$$

where β is the compressibility of the grease and P the pressure at the element. The total excess volume in the pipe, V_x, is

$$V_x = \int_0^L (dV_0 - dV)$$
$$= \int_0^L \pi R^2 \beta P \, dx$$

P varies linearly with x in the following way:

$$P = \frac{(P_e - P_p)x}{L} + P_p$$

Therefore

$$V_x = \frac{\pi R^2 \beta L (P_p + P_e)}{2} \qquad . \quad . \quad (2.1)$$

The effect of the expansion of the pipe by internal pressure

In the above section the pipe has been assumed to be rigid, but conditions may arise in which the distortion of the pipe itself causes a significant increase in V_x. The distortion may be taken into account in a general way by replacing the undistorted internal radius, R, in the integral above by a pressure-dependent radius, R'. The dependence of R' on internal pressure is then obtained from one of the available theories. For example, one of the thick-wall theories leads to the expression

$$R' = R\left(1 + \frac{PR_0}{hE}\right)$$

where R_0 is the outer radius of the pipe, h is the wall thickness, and E is Young's modulus for the pipe material.

The effect of the pressure gradient during the filling and emptying of the pipe is to impose a taper on the pipe. Consequently, the pressure gradient is not linear. A subsidiary calculation shows that for steel pipes of reasonable wall thickness ($R_0/h \approx 8$), the deviation of the gradient from linearity is less than 1 per cent under normal conditions, so the linear distribution of P may be used in the expression for R'.

The integration is then performed to give a new value of excess volume,

$$V_x = \pi R^2 \beta L \left\{ \frac{P_p + P_e}{2} + \frac{R_0}{hE}\left[\frac{(P_p - P_e)^2}{3} + P_p P_e\right]\right\}$$

The relaxation of pressure in a pipe

Consider the case of a rigid pipe. Initially P_p is at 1200 lb/in² and P_e at 600 lb/in² pressure. The pump is then stopped and disconnected from the pipe, when P_p drops to zero. For ease of calculation it will be assumed that P_p attains the value zero instantaneously and that an excess volume V'_x, where

$$V'_x = \frac{\pi R^2 \beta L P_p}{2}$$

extrudes from the pipe instantaneously. This is obviously not true, but of the total time of relaxation the time for this extrusion is small.

Therefore it is assumed that at zero time the pressure gradient is from P_e (600 lb/in²) to zero, and the grease to be extruded has the volume V''_x, where

$$V''_x = \frac{\pi R^2 \beta L P_e}{2}$$

In the case of the deformable pipe

$$V''_x = \pi R^2 \beta L \left(\frac{P_e}{2} + \frac{R_0 P_e^2}{3hE}\right)$$

For typical conditions of $P_e = 600 \text{ lb/in}^2$ ($4.14 \times 10^6 \text{ N/m}^2$), $R_0/h = 8$, and $E = 2 \times 10^{11} \text{ N/m}^2$, the second term in the brackets amounts to only 0·01 per cent of the first and therefore may be neglected in the rest of this calculation.

We can now return to the case of the rigid pipe. During extrusion, P_e and V''_x vary, so that we shall call the initial values P_{e0} and V''_{x0}. In the new notation,

$$V''_{x0} = \frac{\pi R^2 \beta L P_{e0}}{2}$$

and at any instant

$$V''_x = \frac{\pi R^2 \beta L P_e}{2}$$

From Poiseuille's law, for laminar flow of an equivalent Newtonian liquid,

$$\frac{d(V''_{x0} - V''_x)}{dt} = \frac{\pi P_e R^4}{8 \eta_a L}$$

where t is the time of flow and η_a is the apparent viscosity of the grease. Substitution for V''_x and rearrangement gives

$$\frac{dP_e}{P_e} = -\frac{R^2}{4 \eta_a \beta L^2} dt \quad . \quad . \quad (2.2)$$

Case 1

Consider a Newtonian liquid of constant viscosity, η. Integration of equation (2.2) between the limits P_{e0} and P_e gives

$$\ln\left(\frac{P_{e0}}{P_e}\right) = \left(\frac{R^2}{4 \eta \beta L^2}\right) t \quad . \quad . \quad (2.3)$$

In a typical example
$$\beta = 5 \times 10^{-10} \text{ m}^2/\text{N}$$
$$R = 2.5 \text{ cm}$$
$$L = 100 \text{ m}$$
$$\frac{P_{e0}}{P_e} = \frac{600}{200} = 3$$

Then $\quad \eta = 280 t$

Therefore, the material which requires 3 min to relax from 600 lb/in² to 200 lb/in² pressure at the valve end has a viscosity of 5×10^4 poise.

Case 2

Consider a grease with variable viscosity, η_a. Greases of the type represented by the 'average' grease mentioned above can be represented by the following equation, which is a power relationship normalized to a dimensionally consistent form:

$$\frac{\tau}{\tau_0} = \left(\frac{D}{D_0}\right)^n$$

where τ is the shear stress, D is the rate of shear and τ_0, D_0, and n are material constants dependent on temperature and, to a small extent, on pressure. In terms of the apparent viscosity $\eta_a (\equiv \tau/D)$

$$\log \eta_a = \log \eta_0 + (n-1) \log D$$

where $\eta_0 \equiv \tau_0/D_0{}^n$ is the viscosity at a rate of shear of 1 s^{-1}. The linear relationship between $\log \eta_a$ and $\log D$ is valid over a reasonable range of rate of shear at the low rates of shear which, it will be seen, are of interest in this problem. The flow equation may be expressed in terms of η_a and τ as

$$\eta_a = \eta_0{}^{(1/n)} \tau^{(n-1)/n}$$

Substitution in equation (2.2) and replacement of τ by $P_e R/2L$ gives

$$\frac{dP_e}{P_e{}^{(1/n)}} = -\frac{(R/2L)^{(n+1)/n}}{\beta \eta_0{}^{(1/n)}} dt$$

The value of n is always less than one and usually about 0·3, so integration between P_{e0} at time zero and P_e at time t_r gives

$$\frac{n}{n-1} [P_{e0}{}^{(n-1)/n} - P_e{}^{(n-1)/n}] = \frac{(R/2L)^{(n+1)/n}}{\beta \eta_0{}^{(1/n)}} t_r$$

or, in more convenient form,

$$t_r = \frac{n \beta \eta_0{}^{(1/n)}}{(1-n)(R/2L)^{(n+1)/n} P_{e0}{}^{(1-n)/n}} \left[\left(\frac{P_{e0}}{P_e}\right)^{(1-n)/n} - 1\right] \quad . \quad . \quad (2.4)$$

For the 'average' grease in a typical system
$$\eta_0 = 5.8 \times 10^3 \text{ poise } (5.8 \times 10^2 \text{ Ns/m}^2)$$
$$n = 0.32$$
$$\beta = 5 \times 10^{-10} \text{ m}^2/\text{N (approx.)}$$
$$R = 2.5 \text{ cm}$$
$$L = 100 \text{ m}$$
$$P_{e0} = 600 \text{ lb/in}^2 = 4 \times 10^6 \text{ N/m}^2$$

Equation (2.4) becomes

$$t_r = 11.97 \left[\left(\frac{600}{P_e}\right)^{2.125} - 1\right]$$

This function is evaluated in Table 2.1, which shows that the 'average' grease requires about 2 min for relaxation of pressure from 600 to 200 lb/in². It will appear to relax completely in about 20 h. These calculations do not explicitly take into account the yield stress of the grease. It would be expected that pressure would not relax beyond P_0, the value corresponding to the yield stress τ_0 in the formula

$$P_0 = \frac{2L \tau_0}{R}$$

Table 2.1. The variation of P_e with time for the 'average' grease in a 100-m, 5-cm diameter pipe

P_e, lb/in²	$\left(\dfrac{600}{P_e}\right)^{2.125}$	t_r, s
600	1·00	0
500	1·47	6
400	2·36	16
300	4·36	40
200	10·3	110
100	45·0	530
10	6000	72 000

The yield stress of the 'average' grease in pipes is not known, but a value of the order of 10^2 N/m^2 is not unreasonable for grade 2 greases. For a P_e of 200 lb/in^2 the shear stress is 1.7×10^2 N/m^2, so at this critical value of pressure there may already be an effect of yield stress slowing down the relaxation. The corresponding rate of shear (using the logarithmic relationship) is 0.022 s^{-1}. Experimental results obtained at rates of shear as low as this are rare and would seem to be needed.

The time required to compress grease into a pipe

Equation (2.4) gives the time, t_r, for relaxation of pressure to any desired level, P_e. 'Snowballing' will occur if this time exceeds the imposed cycle time, t_c. However, it should be appreciated that even in the absence of 'snowballing' there is a lower limit to the cycle time. With the pump operating continuously the cycle time cannot be less than the time taken to compress grease into the 'relaxed' pipe and to feed the bearings. The possibility exists of having a dispensing system with a relaxation time designed to avoid 'snowballing' but insufficient to allow for the operation of all the valves within the permitted time. There is a shortest possible cycle time, t_{cs}, that can be calculated approximately.

During the available time a volume of grease, V_B, is fed to the bearings and a volume, V'''_x, is compressed into the pipe. For a given pump the rate of delivery, v_p, is known and is normally constant. Then

$$v_p t_{cs} = V_B + V'''_x$$

V'''_x is obtained by modifying equation (2.1) to take account of the value of the end pressure at the completion of the relaxation period, P_{ef}. Then

$$v_p t_{cs} = V_B + \pi R^2 \beta L \frac{P_p + P_{e0} - P_{ef}}{2} \quad (2.5)$$

In this calculation the effects due to branch pipes are neglected.

In a typical example

$$v_p = 10 \text{ oz/min} = 5 \text{ cm}^3/\text{s}$$
$$\beta = 5 \times 10^{-10} \text{ m}^2/\text{N}$$
$$R = 2.5 \text{ cm}$$
$$L = 100 \text{ m}$$
$$P_p = 1200 \text{ lb/in}^2 = 8 \times 10^6 \text{ N/m}^2$$
$$P_{e0} = 600 \text{ lb/in}^2 = 4 \times 10^6 \text{ N/m}^2$$
$$P_{ef} = 200 \text{ lb/in}^2 = 1.3 \times 10^6 \text{ N/m}^2$$

Then

$$t_{cs} = \frac{V_B}{5} + 105 \quad (\text{seconds, with } V_B \text{ in cm}^3)$$

Therefore, with the 'average' grease (cf. Table 2.1), the pump will take longer to complete a cycle than is required for relaxation to 200 lb/in^2 end pressure. The shortest cycle time is thus pump-limited rather than relaxation-limited in this particular example.

DISCUSSION

The cycle time, t_c, is normally composed of the time taken to operate the valves, t_{cs}, and a deliberately imposed rest time, t_{cr}, such that

$$t_c = t_{cs} + t_{cr}$$

In modern plant, lubrication is required at more frequent intervals than hitherto and t_{cr} has been reduced to bring t_c to as low as 3 min. The example worked above shows that even with t_{cr} set equal to zero the delivery of 375 cm^3 of grease on each cycle will take $[(375/5)+105]$ s, or 3 min, of which nearly 2 min are required to compress the grease. There are two consequences of this result. First, the normal method of calculating t_{cs} by dividing V_B (the volume of grease to be delivered to the bearings in a cycle) by v_p (the pump delivery rate) may not be accurate enough. The effect of grease compressibility should be taken into account [equation (2.5)]. Secondly, the observation that a twin-line system of the type considered cannot be successfully operated on cycle times of about 3 min can be explained.

Equation (2.5) cannot by itself be used to re-design the system to reduce t_{cs} because so many of the quantities are interdependent. It would be necessary to calculate the effects of any change in a dimension on pressure drop, flow rate, etc., using pump performance data and the grease flow properties.

If, by a process of adjustments, t_{cs} were reduced to a desirable level, it is possible that the 'snowballing' effect, caused by a too-slow relaxation of pressure, would limit the cycle time. The worked example (Case 2 and Table 2.1) shows that a not-untypical system is close to the limit if a 3-min cycle time is required. A slight drop in temperature, thereby increasing η_0, or a slight increase in pipe length will increase the time for relaxation of pressure, t_r, beyond 3 min, since both η_0 and L are raised to high powers (3.125 and 4.125, respectively, in the example).

Like equation (2.5), equation (2.4) cannot be used directly for the re-designing of a dispensing system, but in conjunction with other available relationships it may be of some assistance.

It has already been emphasized that the calculations have been carried out for somewhat idealized conditions. In practice, conditions will be different and due allowance must then be made for these. Of particular importance will be effects of temperature gradients through the system, of variations in pipe sizes, of elbows, and of the use of compliant pipe materials. There are also uncertainties in the values to be used for the properties of the greases.

The value for compressibility, β, has been taken to be equal to that found for typical lubricating oils, because data are not available for greases. It could be corrected by assuming the gelling component of the grease to be incompressible, in which case the corrected value would be $a\beta'$, where a is the volume fraction of oil and β' is the compressibility of the oil. For greases in centralized lubrication systems a is usually greater than 0.9, so the correction is usually less than 10 per cent. Also, β will decrease slightly as the pressure increases (3 per cent per

1000 lb/in²) and as the temperature decreases (10 per cent in 20 degC). On the whole, the uncertainties in β are not likely to give rise to considerable errors.

The situation is quite different with the viscosity. This changes quite rapidly with the stress available; thus a 10 per cent drop in shear stress for the 'average' grease considered in our examples causes the viscosity to increase by 20 per cent and the flow rate to fall by 30 per cent. The analysis given takes this effect into account, but it is clear that large errors can arise if the flow properties of the grease are not known accurately enough. These estimates for the 'average' grease apply only over the range of validity of the particular power relationship chosen. Now, it is inherent in the problem of pressure relaxation in a pipe that difficulties arise when the stresses available fall to values close to that of the yield stress of the grease. It is here that the parameters of the power relationship are expected to change rapidly with the stress: where n will decrease to zero and τ_0 will become equal to the value of the yield stress. Consequently, long extrapolations of viscometer results by such equations, or any other procedure, should be avoided if possible. The calculations in this paper have shown that data at shear rates lower than $0 \cdot 02$ s^{-1} are required for analysis of the behaviour of the 'average' grease in the 100-m long pipe. The parameters of the power relationship were obtained from measurements down to $0 \cdot 4$ s^{-1}. The extra portion involved here is probably not valid, and the viscosity and the value of t_r (in Table 2.1) for a P_e of 200 lb/in² have been thereby underestimated. Some laboratories possess apparatus for determining grease viscosity by the standard method, A.S.T.M. D.1092. The lower limit of rate of shear with this instrument is about 10 s^{-1}; clearly the errors of extrapolation will be very serious. This limitation may be overcome, at least in part, by fitting larger diameter, non-standard, capillaries to the apparatus. Another solution is to construct special apparatus; for example, the Shell–de Limon rheometer is capable of reaching a rate of shear of $0 \cdot 02$ s^{-1}.

Finally, it is pointed out that for calculations of flow in pipes there is an advantage in using apparent viscosity data determined in pipe flow experiments. If another type of viscometer is used, the data should be adjusted for the different rate of shear profiles existing in the pipe and the viscometer. Failure to make this adjustment can lead to errors of up to about 20 per cent.

APPENDIX 2.1

REFERENCE

(1) CAMERON, A. *Principles of lubrication* 1966 (Longmans, London). Chapter 23 contains an account of rheological properties of greases.

Paper 3

SOME DESIGNS FOR MOUNTINGS INCORPORATING TWO ROLLING BEARINGS AND A GREASE RELIEF SYSTEM

G. W. Mullett[*]

The designs for grease relief systems of the valve type for a mounting containing a single bearing are already well known. In some machinery, however, the bearing requirements cannot be satisfactorily met by a single bearing. In these cases a pair of bearings in the one mounting is needed. The design of the mounting is further complicated if, in addition to pairing the bearings, a grease relief system is also required.

The paper gives eight outline sketches covering various designs of two-bearing grease valve mountings. All have been rig tested and the efficiency of each design categorized. For the experimental work 3-in (76·2-mm) bore medium series ball and roller bearings were used. These were grouped into six pairs. Each pair was tested, either paired in contact or axially spaced apart. Shaft speeds ranged from 1070 to 3600 rev/min, DN = 82 000 to 275 000. In the relief systems, either one or two valve discs were variously positioned above, below, and between the bearings. The paper should be of value to design engineers.

INTRODUCTION

IN A MACHINE a rolling bearing and a grease lubricant are a compatible couple if the mounting has been properly engineered to meet the duty for which it is required. With its first packing of grease the bearing will run trouble-free for a duration ranging from a few hours to as much as 50 000 h, the actual duration depending upon the type of bearing and its environment, notably speed, temperature, and the accuracy of the machine build. Mounting accuracies of inadequate precision are responsible for many premature failures of bearings. The machine-build accuracy in the erection shop is important, but it is more important still to determine what happens to this accuracy when the machine is performing. When running and hot the machine can become flexed and distorted by its working loads and by the differential heat expansions between its various parts. These movements can destroy the as-built accuracy.

A properly greased bearing, once it has settled down, runs on a thin film of grease, the film coating and lubricating its working parts. When this film dries out with time and temperature, or when it is ruptured with time by

The MS. of this paper was received at the Institution on 16th July 1969 and accepted for publication on 22nd September 1969. 24
[*] *Research Manager, Ransome Hoffmann Pollard Ltd, Newark, Notts.*

speed and load, the bearing failure commences. The lubrication life of the grease film has come to an end. This life is not necessarily the same as the life of the grease. Bearings can fail as a result of dryness, with the bulk grease in the mounting still in good condition. The life of the film determines how long the bearing will run. If the film can be repeatedly renewed in good time before the end of film life, failure can be prevented and the service life of the bearing prolonged to many film lives.

The film can be positively renewed by making a fresh start. The mounting is dismantled, the old grease cleaned out, and new grease is packed into place. This method is not always convenient, particularly if the lubrication life is short. It results in the machine being out of service at frequent intervals.

An alternative method is to inject new grease into the mounting in such a manner that it passes through the bearing and creates a new film on its parts. This method is not wholly reliable in practice because it is not always easy to effect properly. Too often greasing nipples are randomly situated on housings, resulting in the injected grease merely filling up the empty spaces in the mounting. If the bearing survives long enough, eventually the bearing is choked with grease. In this state, whilst the bearing is certainly well lubricated, the mounting is virtually pressurized and the grease will force itself out somewhere—

usually through the shaft gland—causing drippage, and the mounting becomes dirty and objectionable. Because the grease is being continually worked, the temperature of the bearing and the power consumed by it are higher than they need be. The extra heat generated hastens the degradation of the grease whilst the extra power is a waste of energy. If grease is injected, an exit for the surplus grease to escape freely from the mounting, and so prevent choking, must be provided. The so-called grease escapage plugs and pipes that are seen on mountings are usually of little value. To move grease through such small apertures, if it can be moved at all, requires a considerable pressure, which can only come from a choked mounting.

A very efficient method of renewing the grease film does exist. It is called the grease valve, and the arrangement was initiated and developed by The Skefko Ball Bearing Company Limited. The method is not new. As far as the author is aware, it was first described in 1947 (1)*. Strangely, the method has not been as widely adopted in practice as one would have expected. For this reason it may be useful to list here the advantages of a grease valve, and so focus attention upon it:

(1) It expels all the surplus grease from a mounting in a clean and orderly manner leaving the bearing filmed, which is all that is required from the grease. Running temperature and power consumption are thus minimums.

(2) Following an injection, the surplus grease can be seen emerging from the eject port in the mounting, therefore the greaser is certain that the system is working properly.

(3) Because new grease can be injected as often as is necessary, the operational speed can be safely increased above that normally used for conventional mountings. The increase is of the order of 30 per cent.

(4) It is impossible to overpack and choke the mounting. An over-enthusiastic greaser can do no harm, only good; but he does waste grease, a minor point. Grease is cheap but down-time due to failure is expensive.

(5) It enables a bearing to run as long as the overhaul period of the machine; indeed, it could run as long as the life of the machine. To regrease the bearing it is not necessary to stop the machine, which can continue performing its duty without check.

(6) It enables a bearing which operates under really arduous conditions to give a service life of useful, practical value. For example, by injecting new grease three or four times a day, a bearing can be run at temperatures as high as 250–300°C for a few hundred hours.

This paper is not concerned with single-bearing grease valve mountings. They are adequately described elsewhere (1). It is concerned with double-bearing valve mountings, about which there appears to be a general lack of information.

In some machinery, particularly those machines of the larger sizes, the requirements of loads, speeds, and bearing sizes can be such that they cannot be adequately accommodated by single bearings. It is then necessary to fit two bearings, side by side, in the one mounting. One such arrangement comprises a ball bearing to react the thrust load, and a roller bearing to react the journal load; in this manner the loading is shared by two bearings. Another common arrangement is a preloaded pair of ball bearings used when it is necessary to restrict the displacement of a loaded shaft.

If the conditions of such two-bearing mountings are at all arduous (the very fact that there are two bearings tends to make them so), an insurance against lubrication failures is required. The best form of insurance is the valve type of grease relief system. The problem then arises of how to design an efficient system into the double mounting, and it is with this problem that this paper is concerned.

SCOPE OF THE PAPER

The paper outlines the test programme which has been carried out. The end-product of the work done is given in the form of eight line-sketches, Figs 3.1–3.8. Each figure is labelled Set-up 1 to 8 and suitably annotated. Each figure depicts a two-bearing grease valve mounting for one end of a vertical shaft.

These eight set-ups were built and subjected to grease injection tests and their efficiencies noted. The set-up was classed efficient when each successive injection of new grease resulted in only a temporary increase in bearing temperature and a similar amount of grease to that injected was ejected by the valve.

The work occupied some 20 months and 8000 running hours. Sufficient will be written in this paper to outline the programme, but it is not necessary to detail the individual tests.

The paper is intended for the engineering designer. It is hoped that the sketches and the observations made will assist him in his work.

THE TEST BEARINGS

Eight types of bearings were used, from the medium series; boundary sizes being the same for each, namely, 3 in (76·2 mm) bore × 7 in (177·8 mm) outside diameter × 1·5625 in (39·7 mm) wide. They are listed in Table 3.1.

The question of the maximum speeds at which bearings can be run, using a grease lubricant, is a controversial matter. Most bearing makers do list in their catalogues their considered maximum speed values, but they are not too specific in defining all the requirements which must be met by the user for the speeds to be satisfactorily attained in practice. Perhaps this is not surprising when it is considered that the performance of a bearing depends more upon how and for what purpose and by whom it is used than upon the bearing itself. The maximum speed which can be recommended for a bearing is thus a matter for judgement, taking all the circumstances into account. Manufacturers' speeds must be regarded with an enquiring mind and not accepted too readily at their face

The reference is given in Appendix 3.1.

Table 3.1. The eight test bearings

R. & M. designation	Description
MJ 3 LOC	Rigid single-row ball journal (upper bearing in Fig. 3.7).
MJT 3	Rigid single-row ball 20° angular contact (Fig. 3.3).
DMJT 3(LOC)	Rigid single-row ball 35° angular contact, a duplex (four-tracked) bearing (upper bearing in Fig. 3.1).
MRJ 3	Rigid single-row roller journal (lower bearing in Fig. 3.1).
MJ 3M LOC MJT 3M DMJT 3M(LOC) MRJ 3M	These four bearings, suffix M, are the higher speed equivalents of the above four. They are fitted with machined brass cages centred on the rolling elements.

LOC means location. A location bearing is 0·004–0·007 in undersize on its outside diameter. Since it is slack in a standard-size housing, when paired with a roller journal bearing (as in Fig. 3.1), it cannot withstand a journal load and reacts a thrust load only. Alternatively, a standard bearing can be used, e.g. MJ 3 and not MJ 3 LOC, the housing itself being bored 0·004–0·007 in oversize.

values. Bearings can certainly be run at quite high speeds in grease, but such speeds have little practical value if the lubrication life is only a matter of a few hours.

The speeds given in Table 3.2 are the author's opinions of the recommended maximums. They apply to single bearings in conventional mountings in machines built to a good standard of engineering. Using a quality grease, the lubrication life should be of the order of 1500–2000 h provided the running temperature does not exceed some 50°C and the loading is moderate.

The eight types of bearings can be grouped into six pairs for double-bearing mounting. These pairs are set out in Table 3.3. The grouping arises from practical considerations. As the maximum speed allowable for a pair is based on the speed of the bearing which has the lower maximum, there is no point in pairing a higher speed bearing with a lower speed bearing; in addition, each pair

Table 3.2. Single bearing recommended maximum grease speeds

R. & M. designation	Speed, rev/min	R. & M. designation	Speed, rev/min
MJ 3	3250	MJ 3M	3550
MJT 3	1400	MJT 3M	2500
DMJT 3	1575	DMJT 3M	3150
MRJ 3	2550	MRJ 3M	3500

selected must be capable of withstanding a thrust load in either direction, the bearing in the other mounting, at the other end of the shaft, being most likely a roller journal bearing.

Table 3.3 also lists the author's opinion of the recommended maximum speed for each pair in a grease relief mounting, subject to the same stipulations as the speeds of Table 3.2 with the altered requirements of a grease injection every 300 h and a running temperature not exceeding 70°C.

The injection period of 300 h is about 50–70 per cent of the lubrication life which could be expected from the paired bearings if they had been fitted in a conventional mounting. This is 'playing safe'. The lubrication life to the point in time when a bearing becomes a complete failure is of no real practical value. Some types of bearings will run outwardly apparently all in order but inwardly in an advanced state of failure. It is of practical value to know the time a bearing can be run before the lubrication begins to deteriorate. For safety, the injection period should, preferably, not exceed 50 per cent of this time.

THE TEST MOUNTINGS

Two bearings can be arranged in a single mounting in two ways, either in actual contact or with a short axial space between them. The space is usually 60–70 per cent of the width of one bearing. Whilst spacing them apart does increase the angular rigidity of the mounting, the more

Table 3.3. The six pairs of test bearings

No.	Bearing pair	Recommended max. speed, rev/min	Remarks
1	MJT 3 MJT 3	1175 1375	The slower speed for each pair is permissible when the two bearings are touching, as in Fig. 3.1. The higher speed applies when the bearings are spaced and divided, as in Figs 3.5 and 3.6. When spaced but not divided, as in Fig. 3.4, take the average speed.
2	MJ 3LOC MRJ 3	1850 2500	
3	DMJT 3(LOC) MRJ 3	1125 1550	
4	MJT 3M MJT 3M	2150 2450	
5	MJ 3MLOC MRJ 3M	2500 3400	
6	DMJT 3M(LOC) MRJ 3M	2250 3075	

important aspect is that it permits other pieces of the mounting to be placed between and so divide the bearings. In this manner, each bearing can be isolated from the other and each can perform in much the same way as a single bearing without interference from the other. This aspect is reflected in the higher maximum speeds for the spaced apart and divided mountings, Table 3.3.

Either a single valve disc or two valve discs can be used. In the former case the disc can be positioned either above or below or between the bearings, as in Figs 3.1, 3.3, and 3.5, respectively. With two discs, one can be positioned above the pair and the other below, Fig. 3.6, or one above each bearing, Fig. 3.7, or one below each bearing, Fig. 3.8.

For each disc there must be a grease injection point positioned at the face of the bearing(s) opposite to the disc.

All these practical considerations lead to eight possible set-ups.

The figures show various forms of shaft sealing and certain bearing pairs. This is for illustrative purposes only. The sealing methods and the bearing pairs shown can be applied to each of the eight set-ups.

THE TESTS

To cover the programme completely it would be necessary to test each of the six pairs of bearings of Table 3.3 in each of the eight set-ups, Figs 3.1–3.8, i.e. 48 tests. It transpired that such a large undertaking was not needed. It was found that if a set-up functioned efficiently at a shaft speed of N rev/min, then it remained efficient at speeds lower than N rev/min, but it was not necessarily efficient at speeds higher than N rev/min. In actual fact only 18 tests were carried out. Six of these used Pair 5 bearings, the pair with the highest recommended speed, 3400 rev/min, Table 3.3. These six tests were conducted at the slight overspeed of 3600 rev/min, $DN = 275\,000$,

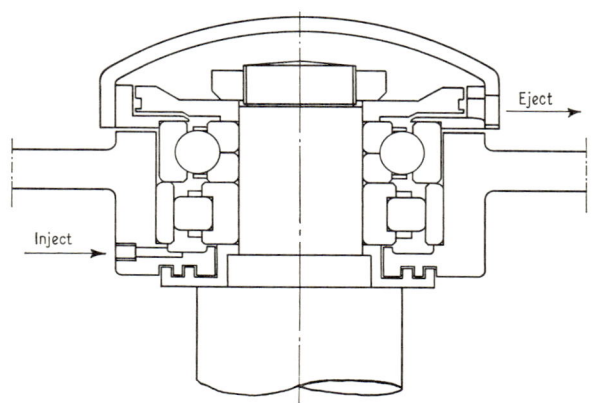

Set-up 1. This arrangement is not wholly acceptable. It should not be used unless the speed is low, below $DN = 80\,000$.

Fig. 3.1

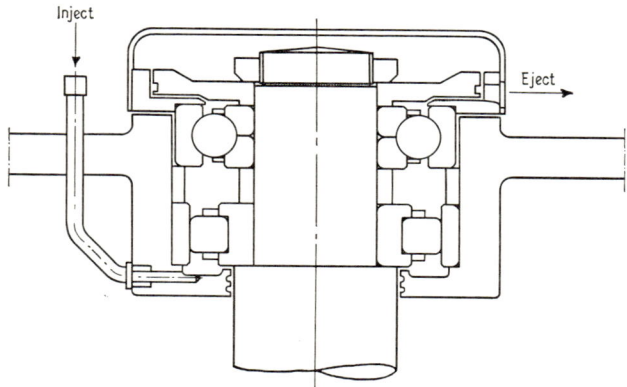

Set-up 2. This arrangement should never be used.

Fig. 3.2

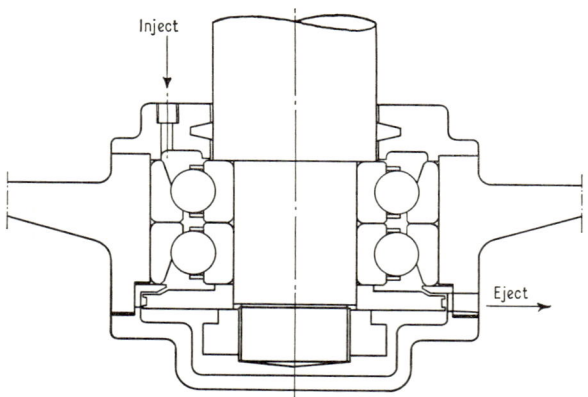

Set-up 3. This arrangement is efficient and suitable for very high speeds up to $DN = 275\,000$.

Fig. 3.3

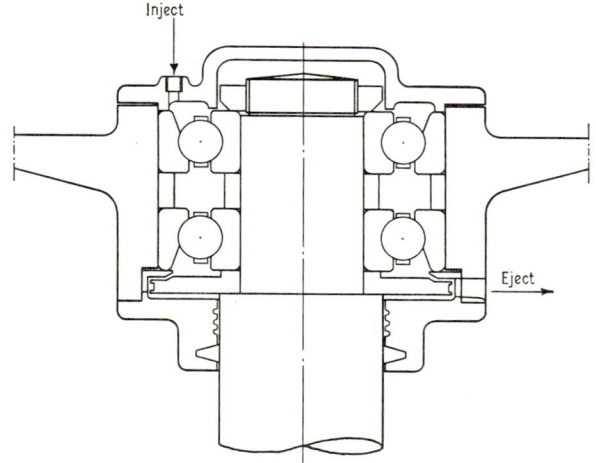

Set-up 4. This arrangement is reasonably efficient and suitable for high speeds up to $DN = 220\,000$.

Fig. 3.4

D is the bearing bore in mm, certainly a creditable performance. Test speeds ranged from 1070 rev/min, Pair 3, to 3600 rev/min, Pair 5.

In each test both bearings were initially packed full of grease, flush with their faces. In addition, the trays or caps into which the new grease was injected were also initially packed full of grease. The lubricant used was a good quality N.L.G.I. No. 3 lithium base grease.

Injections were made twice in every 24 h of running using screw-down Stauffer pots. Such a frequency as this would not be necessary in practice, of course. Each injection measured $1\frac{1}{4}$ oz (35 g) of grease.

In the four set-ups, Figs 3.1–3.4, the injected grease charge was required to pass through both bearings whilst in the remaining set-ups, Figs 3.5–3.8, each bearing received its own charge of $1\frac{1}{4}$ oz (35 g).

The settled running temperatures, measured on the housings, lay between 50 and 60°C.

SOME OBSERVATIONS

The size of a valve disc is important. The higher the speed the larger it should be, and its outside diameter should exceed the outside diameter of the bearing. If the disc is placed below a bearing, it can be slightly smaller than a disc which is placed above a bearing. In the tests, upper discs were 8 in (203 mm) and lower discs 7·4 in (188 mm) diameter, being respectively 1·14 and 1·06 times the bearing outside diameter, which was 7 in (178 mm).

Grease migrates downwards within a mounting more easily than upwards. It is therefore preferable to place a disc below a bearing rather than above it.

The grease eject ports are shown in section in Figs 3.1–3.8. In the other view, their circumferential lengths should not be stinted, and a length equal to about 1·3 times the bore size of the bearing is recommended. In the tests the port length was 4 in (102 mm).

No restrictions whatsoever should be so placed that they hinder the free passage of the ejected grease. This passage to atmosphere should be as short as possible. If the eject passage must have some length (in a radial direction), then its cross-sectional area should increase as the passage is traversed. It is usual to place a sheet metal

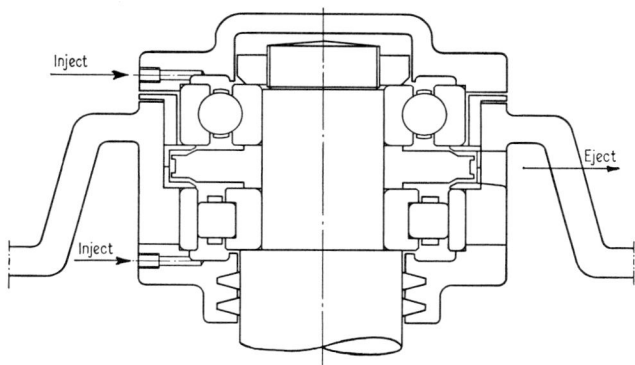

Set-up 5. This arrangement is reasonably efficient and suitable for high speeds up to $DN = 220\,000$.

Fig. 3.5

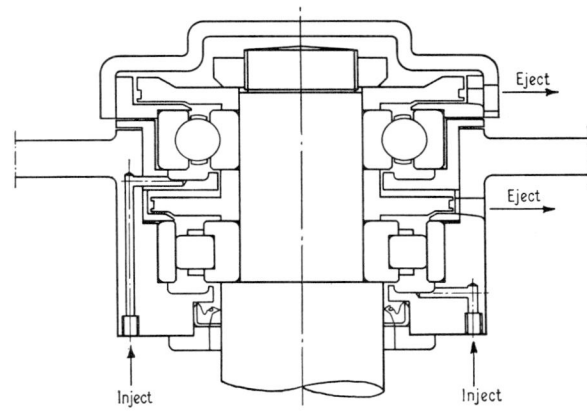

Set-up 7. This arrangement is not wholly acceptable. It should be used for moderate speeds only up to $DN = 150\,000$.

Fig. 3.7

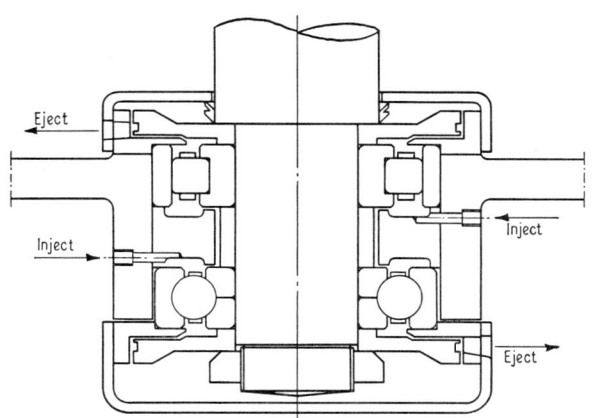

Set-up 6. This arrangement is reasonably efficient and suitable for high speeds up to $DN = 220\,000$.

Fig. 3.6

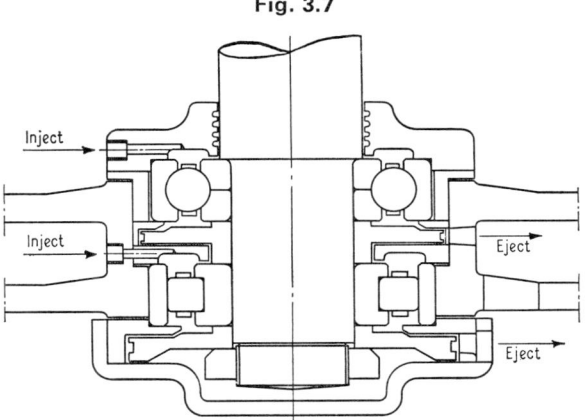

Set-up 8. This arrangement is efficient and suitable for very high speeds up to $DN = 275\,000$.

Fig. 3.8

Table 3.4. Recommendations

Set-up No.	Speeds below DN*			
	Low 80 000	Moderate 150 000	High 220 000	Very high 275 000
1	✓	×	×	×
2	×	×	×	×
3	✓	✓	✓	✓
4	✓	✓	✓	×
5	✓	✓	✓	×
6	✓	✓	×	×
7	✓	✓	×	×
8	✓	✓	✓	✓

* D = bearing bore, mm; N = shaft speed, rev/min.

In a mounting the maximum speed permissible is decided either by (1) the DN maximum for the bearing pair, or (2) the DN maximum for the set-up, whichever is the lower.

catcher below the point of ejection to atmosphere, at the end of the passage. This catcher must be cleaned out regularly.

The running clearance between the disc and the adjacent stationary parts should range from 0·06 in (1·5 mm) in small bearings to 0·12 in (3 mm) in large bearings. For the tests, 3-in bore bearings, the clearance was 0·08 in (2 mm).

The size of each inject G g can be determined from the formula (1),

$$G = \frac{BW}{210}$$

where B is the bearing outside diameter (mm) and W is the bearing width (mm).

For the 3-in bearings tested the formula evaluates at 34 g (1·2 oz) per bearing.

Grease is cheap. Do not take risks by trying to economize either by reducing the quantity injected or by stretching out the relubrication intervals.

Injections should preferably be made whilst the bearings are running rather than stationary. Where a mounting has two inject points it is preferable to make the second a few hours after the first. These are minor points, however, and the choice is open.

All the tests were made with a vertical shaft. The author is confident that the set-ups will function equally well with a horizontal shaft, with one stipulation—the grease eject ports must all point vertically downwards.

CONCLUSIONS

The conclusions are the remarks beneath each of Figs 3.1–3.8. The tests rejected Set-up 2, Fig. 3.2. Table 3.4 gives a summary of the findings as a ready reference for designers.

ACKNOWLEDGEMENTS

The author is indebted to Ransome & Marles Bearing Co. Ltd for permission to publish this paper and to the staff of the Research and Development Department for their skill in preparing the rigs and conducting the tests.

APPENDIX 3.1
REFERENCE

(1) ANON. 'Ways and means of lubricating SKF rolling bearings', *Ball Bearing J.* 1947, Nos 3 and 4 (The Skefko Ball Bearing Co. Ltd, Luton).

Paper 4

IP DYNAMIC ANTI-RUST TEST, LUBRICATING GREASES

F. E. H. Spicer*

The Institute of Petroleum has recently published a test method (IP 220/67) for assessing the rust-prevention characteristics of greases. Specially selected and preserved double-row, self-aligning ball bearings are used as test specimens.

The test, which lasts for $164\frac{1}{2}$ h, is dynamic, the grease film being produced and part of the test being performed with the bearings rotating partially submerged in distilled water. The performance of the grease is assessed by estimating the area of rust on the outer races of the test bearings.

The precision of the test method was established by an extensive international correlation programme involving 30 laboratories from five European countries. This showed a precision which varied according to the performance level of the test greases. The results of tests with greases affording very good protection and with greases affording very poor protection against rusting gave better precision than results from greases with intermediate performance levels.

INTRODUCTION

THE DETRIMENTAL EFFECTS of large quantities of rust in rolling bearings needs no emphasis. Corrosion products occupy the small clearances essential to the bearing operation, and the mechanism seizes. With smaller quantities of rust a bearing will rotate, but heavy wear may occur as a result of abrasion by the rust particles. The highly finished surfaces essential for smooth operation of the bearing are damaged, and vibration occurs. Even minute areas of rust, barely visible to the naked eye, will provide the starting point for pitting or flaking which will seriously reduce the operational life of rolling bearings.

The Institute of Petroleum (IP) Standardization Sub-Committee D concluded in 1958 that there was a need for a standard test method to determine the rust-preventing properties of grease for use in rolling bearings. The Mechanical Tests (Grease) Panel, ST-D-3, was given the task of preparing the method.

PRELIMINARY CONSIDERATIONS

Type of test

Panel ST-D-3 is concerned only with dynamic tests involving rolling bearings, and it might be thought that such conditions were not essential to a rust-prevention test for grease. Indeed, many static tests were already in use. However, the literature [1]† indicates that in some circumstances the movement of corrosive materials and of inhibitors has an effect on the progress of corrosion, which suggests that a dynamic test is desirable. It has also been shown that, irrespective of the type of test selected, the distribution of the grease on the test specimen is of major importance. In some tests, great care is taken to prepare an evenly distributed continuous film of grease on the variously shaped specimens. In spite of this, the great variety of grease consistency, texture adhesive and cohesive properties, and the difficulty of moving grease without entraining air considerably reduce the repeatability of this approach. Even if a perfect film were obtainable, its use in a test method could not be justified, because in practice it is never achieved. The rust-preventing properties of grease must be assessed on the assumption that there will be discontinuities in the grease film.

In the particular case of assessing grease for rolling bearing applications, the most convenient and practicable way of distributing the grease is to place it in a suitable bearing, which is then rotated.

Test specimen

The constituents and the surface finish of materials are

The MS. of this paper was received at the Institution on 17th September 1969 and accepted for publication on 21st November 1969. 33
* *Chairman, Mechanical Tests (Grease) Panel, Institute of Petroleum, 61 New Cavendish Street, London, W.1.*
† *The reference is given in Appendix 4.3.*

known to have an important influence on their susceptibility to rusting. Therefore it was essential to use test pieces of rolling bearing materials machined to the same surface roughness as a rolling bearing. The obvious choice would be a suitable bearing from a bearing manufacturer's normal production range. Such a test piece would be readily available, and the high standard of manufacturing control and inspection would ensure uniformity.

Available test methods

With these basic requirements in mind the panel considered suitable tests which were already in use. Of these, the two most promising candidates were the Co-ordinating Research Council (CRC) Method L-41-1957 and the Emcor test developed by SKF, Gothenburg.

The CRC method had been developed for and correlated with the service performance of greases used to lubricate aircraft landing-wheel bearings. In the test, Timken taper roller bearings are run for a short period under dry conditions, then stored for two weeks in a 100 per cent relative humidity environment.

The panel considered the Emcor test, in which the bearings are run and stored under wet conditions, to be more typical of general wet operating conditions in industry. This test was therefore selected as a basis for an IP method. The test conditions finally agreed upon for the IP method are given in Appendix 4.1, and they are only slightly different from those of the original Emcor test.

Test bearing

The bearing used as the test specimen is a standard production double-row, self-aligning ball bearing. To achieve improved test precision the bearings are specially selected and packed. These bearings may be dismantled, cleaned, and inspected with comparative ease. In the initial test programme the bearings were, in fact, completely dismantled before the test. However, the final form of the test method does not require that this should be done, as it was considered desirable to minimize the handling of the bearings.

Environment

In the Emcor test the bearing rotates partially immersed in distilled water at normal room temperature, and these conditions were adopted for the IP test method. At first it was considered that the carbon dioxide content of the water should be standardized by boiling and cooling immediately before the test. However, at a later stage it was considered that pipetting the water into the bearing housing would reintroduce carbon dioxide; therefore this treatment was omitted.

Suggestions that the test could be accelerated by using salt solution were rejected, as this would orientate the test towards special conditions and would probably reduce its sensitivity.

Duration

The duration of the test is $164\frac{1}{2}$ h. After an initial $\frac{1}{2}$ h of rotation to distribute the grease in the bearing, water is added. Next, the bearing is rotated for 8 h in each of the first three 24-h periods; it then remains stationary for the final 108 h. Thus, both static protection and dynamic rust protection, in the presence of free water, are required from the grease.

Speed of rotation

During the dynamic parts of the test the bearing rotates at 80 rev/min. This comparatively low speed is adequate to ensure grease distribution and the mixing of the water with the grease within the bearing. By using a low speed, sealing and evaporation difficulties are minimized.

Load

The test was not intended to assess the load-carrying properties of the grease under wet operating conditions, and no bearing loading system is employed. However, the design of the rig resulted in the bearings being lightly loaded, due to the weight of the shaft and to slight misalignment of the bearing housings within the manufacturing tolerances. These loads are considered to be of no significance under the test conditions.

The test rig

A description of the rig in its final form is contained in IP Method 220/67 (see Appendix 4.1). It comprises a shaft on which may be mounted up to eight test bearings. The bearings are housed in simple plummer blocks with rubber lip seals to minimize evaporation losses.

Two types of rig are available. In one type the plummer blocks are made of cast iron, the shaft of stainless steel, and the shaft adaptors of mild steel; in the other, a more recent type of rig, the only exposed metal parts are the test bearings. All the other parts are either made from or coated with plastic.

Although the rigs with the cast-iron plummer blocks require more grease, the correlation programme indicated no significant difference in the performance of the two types of rig. The plastics rig is therefore preferred because it is considerably easier to clean and prepare.

Initial precision evaluation programme

Eight members of the panel made duplicate tests on 14 greases in the initial programme. The bearings were completely dismantled before and after the tests and each element of the bearing was assessed. The greases used included calcium, sodium, sodium/calcium, and lithium base materials.

The precision of the method assessed by this programme was disappointing, but there were indications that it could be improved by making slight modifications to the test procedure.

It was clear that there were variations in the quantity of water in the plummer blocks at the end of the test period. This was explained by different air-flow rates, and therefore different evaporation losses from bearing to bearing. To reduce this effect, Vee-ring seals were fitted to the shaft.

Work carried out by the Swedish Standards Commission while the initial IP programme was in progress revealed that the preservative fluid on the test bearings was not always completely removed by the washing techniques then used. Therefore changes were made in the protective fluids, washing solvents, and method of washing the bearings. In addition to these and other minor changes in the method, it was considered that the precision would be improved by a change in the rating system. It was decided that only the outer race should be assessed so that a rating system could then be based on the area corroded.

International precision evaluation programme

Because of interest shown by other European standardization organizations a second precision evaluation programme was arranged in which they were invited to participate. The geographical distribution of the laboratories was as follows:

France	3 laboratories
Germany	6 laboratories
Holland	3 laboratories
Sweden	9 laboratories
U.K. (IP members)	9 laboratories

To ensure that the results obtained would be suitable for statistical analysis and that the precision of the method could then be given in accordance with Appendix E of the IP Standards, a properly designed co-operative programme was undertaken.

Table 4.1. Greases used in international precision evaluation programme

Grease	Thickener	Oil type	Rust inhibitor
A	Ca soap (acid stabilized)	Mineral	None
B	Ca soap (water stabilized)	Mineral	None
C	Na soap	Mineral	None
D	Li soap	Mineral	Non-water-soluble
E	Li soap	Mineral	Water-soluble
F	Li soap	Mineral	None
G	Clay	Mineral	None
H	Li soap	Diester	None

Eight greases were used, details of which are given in Table 4.1. These greases typified the major proportion of greases commercially available. Each laboratory made two separate tests on each grease with a random assignment of the bearing position to be used.

Among the 30 participating laboratories, 32 test rigs were available, 8 with cast-iron bearing housings and 24 with plastics bearing housings.

The tests were conducted and the results rated strictly in accordance with the method given in Appendix 4.1. Results of the tests are given in Table 4.2.

These results were analysed by the IP Precision Evaluation Panel, whose report is given in Appendix 4.2. The main points from this report are:

(a) There was no variation in the results that could be attributed to the location of the laboratory.

(b) The rigs with cast-iron bearing housings gave

Table 4.2. Results of international precision evaluation tests

Grease	Laboratory and type of rig*
	2 3 4 5 6 7 8 8 9 10 11 12 13 14 15 16 17 18 19 20 21 22 23 24 25 26 27 28 29 30 31
	P P C C C C C P P P P P P P P P P P P P P C C P P P P P P P P C
A	3 0 1 3 2 2 2 2 1 0 2 2 2 2 0 2 2 2 3 2 0 2 2 3 2 3 2 2 2 2 3 2
	3 0 2 2 2 2 2 2 2 0 0 2 2 3 0 2 2 3 3 2 0 2 3 2 3 2 0 2 2 2 3 2
B	3 3 3 3 4 3 3 3 3 3 4 3 4 3 3 3 3 3 3 3 3 3 3 4 3 3 3 4 3 4 3
	3 3 3 3 4 3 3 3 3 3 3 4 3 4 3 3 3 3 3 3 2 3 3 3 3 4 3 3 3 3 4 3
C	1 0 1 0 0 1 0 0 0 0 0 0 0 0 1 1 1 2 0 0 0 0 0 0 0 2 0 0 0 0 0
	1 0 0 0 0 1 0 0 0 0 0 0 0 0 1 1 1 2 0 0 0 1 0 1 0 0 0 0 0 0 0
D	1 0 0 2 0 0 0 0 0 0 0 0 0 0 0 0 0 0 0 1 0 0 0 0 0 0 0 1 0 0
	1 0 0 1 0 1 0 0 0 0 0 0 0 0 0 0 0 0 0 0 2 0 0 1 0 0 0 1 0 0
E	0 0 1 0 0 0 0 0 1 0 0 0 0 0 0 0 0 0 0 0 0 0 0 0 2 0 3 1 0 0
	1 0 0 0 0 0 0 0 1 0 0 0 0 0 0 0 0 0 0 0 1 0 0 0 0 0 0 1 1 0 0
F	4 4 4 3 3 4 5 4 2 3 5 3 3 3 4 4 3 2 4 4 4 4 4 2 5 4 4 3 5 5
	4 3 3 3 3 4 5 4 4 3 3 5 4 3 4 4 3 3 2 4 4 4 3 4 4 0 5 4 2 4 5 5
G	5 5 5 5 5 5 5 5 5 5 5 5 5 5 5 5 5 5 5 5 5 5 5 5 5 5 5 5 5 5 5
	5 5 5 5 5 5 5 5 5 5 5 5 5 5 5 5 5 5 5 5 5 5 5 5 5 5 5 5 5 5 5
H	1 0 0 0 0 1 0 0 0 0 0 0 0 0 1 1 0 0 0 0 0 0 2 0 0 1 1 0 0 0 0
	1 0 1 0 0 0 0 0 0 0 0 0 0 0 1 1 1 0 0 0 1 2 1 0 1 0 0 0 0 0 0

* C = rig with cast-iron plummer block; P = rig with plastics plummer block.

slightly higher rating figures than those with plastics housings, but the difference was not significant statistically.

(*c*) The precision of the result varied according to its value.

Greases which gave very good or very poor protection against rust gave results with good reproducibility. At intermediate levels of protection results were less reproducible. For this reason the precision cannot be expressed as a single figure and is given in Table 4.4 (Appendix 4.1).

Table 4.4 shows that for results obtained under repeatability conditions as defined in Appendix E of the IP Standards, the test is suspect if pairs of determinations vary by more than one rating level or if determinations rated 0 to 5 show any variation.

For results obtained under reproducibility conditions the test is suspect if duplicate determinations vary by more than two rating levels, or if pairs of determinations which include ratings of 0 or 5 vary by more than one rating level.

Relevance of the method

The difficulties encountered in proving a relationship between a given test method and even a restricted service application are well known to anyone engaged in test development. Therefore it will be appreciated that it is quite impracticable to carry out a correlation programme for a test with such a wide sphere of application as the IP 220/67 Method. Nevertheless, the selection of the test specimen and conditions gives good reason for believing that the test will indicate the likely performance of greases in service. This belief is supported by the fact that greases E and G were supplied as commercial types, E being typical of grease giving good rust protection and G of grease giving poor rust protection.

The test could quite easily be adapted to assess the performance of greases in corrosive environments other than that described in the method. However, by doing this the precision of the method would almost certainly be affected. It cannot be over-emphasized, therefore, that the precision quoted for this test method, as with all IP standard tests, is only applicable where the test has been performed exactly as described in the method.

ACKNOWLEDGEMENT

The author wishes to thank the Council of the Institute of Petroleum for permission to publish this paper.

APPENDIX 4.1
IP DYNAMIC ANTI-RUST TEST—LUBRICATING GREASES
IP 220/67

Introduction

A dynamic anti-rust test has been developed for general evaluation of the rust-resisting performance of a lubricating grease. Eight test bearings can be used simultaneously, but only two of these bearings are necessary to run the rig.

Scope

The method is intended to evaluate lubricating greases with respect to their rust-prevention characteristics.

Outline of method

The grease is tested in a ball bearing running at 80 rev/min under no applied load in the presence of water. The grease is evaluated by examining the outer race of the bearing, after test, for freedom from rust.

Apparatus

Test rig—power operated with each bearing mounted on a single shaft. The rig may consist of standard iron plummer blocks and steel shaft, or rig parts may be made from a polyamide plastic. A suitable rig is detailed below under 'Additional specifications'.

Test bearing—a double-row self-aligning ball bearing (30 × 72 × 19 mm), SKF 1306 K/236 725, specially inspected and packed to eliminate rust, shall be used.

Materials

White spirit—distillation range 150–200°C.

Lint-free cloth.

Distilled water.

Protective gloves—smooth clean p.v.c. or polythene.

Solvent rinse solution—isopropyl alcohol, 90 per cent vol.; distilled water, 9 per cent vol.; ammonium hydroxide (sp. gr. 0·880), 1 per cent vol.

Preparation of apparatus

(*a*) Remove, by wiping, all traces of grease from previous test, from plummer blocks. Wash the plummer blocks in the solvent rinse solution followed by distilled water. Dry thoroughly, using the lint-free cloth.

(*b*) Two new bearings shall be used for each test and shall not be touched with the fingers at any time.

(*c*) Number bearings on the outside diameter of the outer ring. The easiest way of doing this is to use an electric pen. Acid etching must not be used.

Note 1. Protective gloves shall be worn for operations (*d*)–(*k*).

(*d*) Wash the bearings thoroughly in hot, 125–150°F (50–65°C), white spirit to remove the rust preventive. Repeat the wash using fresh hot white spirit to ensure complete removal of the rust preventive.

(*e*) Transfer the bearings from the white spirit to the solvent rinse solution to remove any white spirit which may be present. Then rinse the bearings and rotate slowly in freshly made solvent rinse solution heated to a minimum of 150°F (65°C).

Caution

The washing temperatures specified are considerably above the flash point of the solvent. Accordingly, the washing operations should be carried out in a well-ventilated hood where no flames or other ignition sources are present.

(f) Remove the bearings from the solvent rinse solution and place on a filter paper to drain. After draining, dry the bearings in an oven at 160°F (70°C) for 15–30 min.

(g) Permit the bearings to cool to room temperature and re-examine the surfaces to ensure that corrosion-free and free-turning specimens have been selected. (Care should be taken not to spin the bearings after cleaning and drying.)

(h) Inspect the outer ring tracks using a dentist's mirror (no magnification). If etch spots or corrosion are evident, reject the bearing.

(i) Distribute 10 ± 0.1 g of the grease evenly in each test bearing using a spatula. Rotate the outer ring slightly by hand to assist distribution. Smear all the external surfaces of the bearing with the grease.

(j) If the all-metal rig is used, then an additional quantity of the test grease shall be filled in the lower half

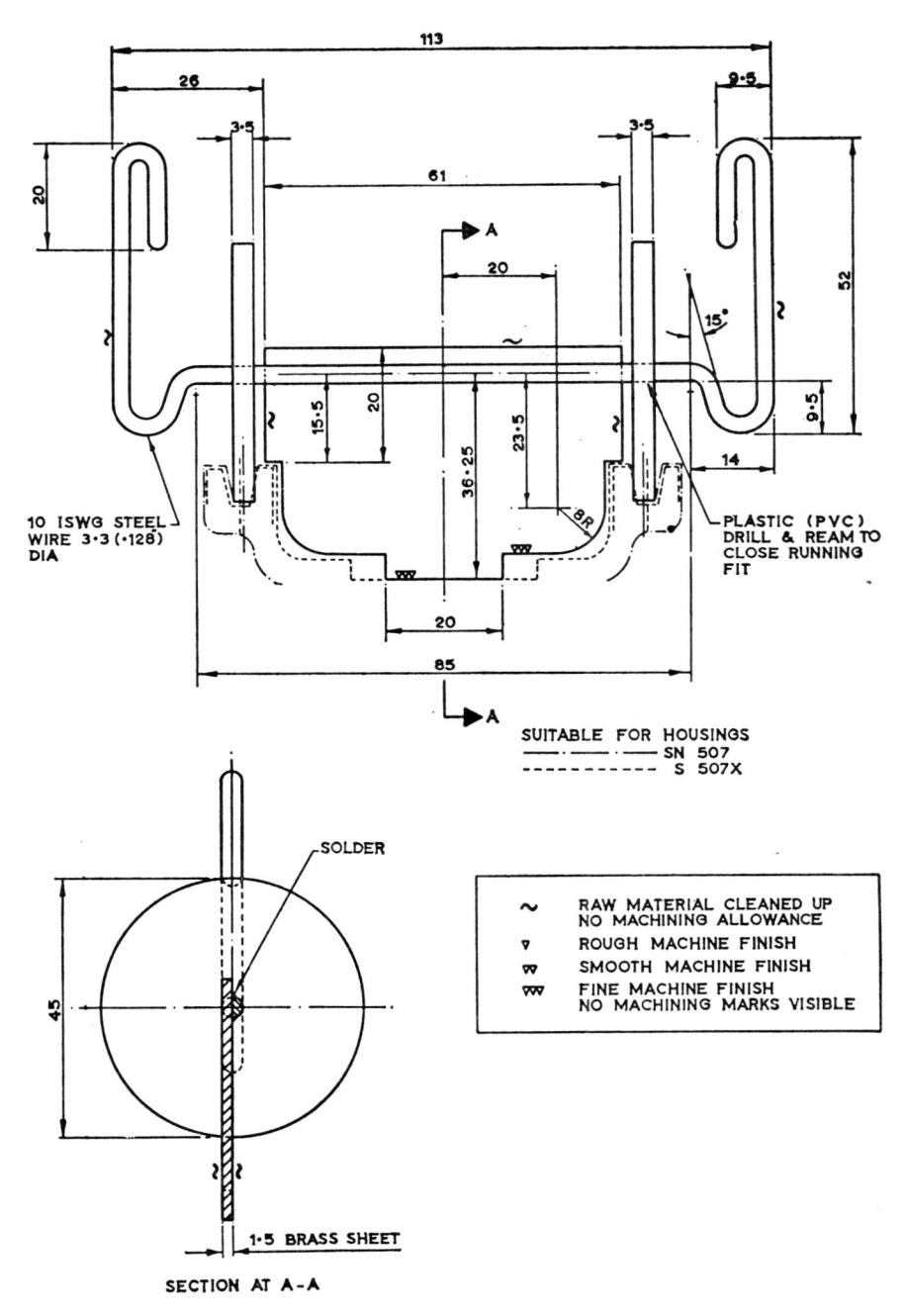

All dimensions in millimetres

Fig. 4.1. Form template for grease pack when all-metal rig is used

of the plummer blocks, using the template shown in Fig. 4.1. No grease is placed in the polyamide plastic plummer blocks.

(*k*) Place the adaptor sleeves, bearings, and Vee-ring seals in position on the shaft and finger-tighten the sleeve nuts (in the case of the all-metal rigs a spanner shall be used). The operation shall be carried out with the shaft suitably supported on the work bench.

(*l*) Place the shaft complete with greased bearings in position in the rig, care being taken that the bearings are central in the plummer blocks.

(*m*) Place the top halves of the plummer blocks in position and finger-tighten the locking screws.

(*n*) Press the Vee-ring seals up against the plummer blocks using the special tool.

Procedure

Make duplicate determinations.

(*a*) Run the rig for 30 min at 80±5 rev/min immediately after assembly to distribute the grease evenly.

(*b*) Remove the top halves of the plummer blocks and introduce 10 ml distilled water into each side of each plummer block using a pipette (i.e. a total of 20 ml). Refit the top halves of the plummer blocks and screw down finger-tight.

(*c*) Run rig 8 h, stop rig, allow to stand for 16 h; run rig for 8 h, stop rig, allow to stand for 16 h; run rig for 8 h, stop rig, allow to stand for 108±2 h.

Note 2. 'Procedure' operations (*b*) and (*c*) must be carried out with a minimum of delay.

Dismantling the apparatus

(*a*) Remove the top halves of the plummer blocks. Lift the shaft and bearings on to a suitable support on the work bench.

(*b*) Remove the bearings and Vee-ring seals from the shaft in the following manner:

   (1) Unscrew the sleeve nut one or two revolutions.
   (2) Tap the end face of the bearing lightly, using a hammer and drift to free bearing.
   (3) Pull bearing, seals, and sleeve off the shaft.

(*c*) Swing the outer ring of the bearing out and prise the balls out of the cage pockets, thus allowing the bearing to be dismantled.

(*d*) Wash the outer ring of the bearing in the cleaning solvent rinse solution and dry, using a lint-free cloth. The bearing is now ready for examination, which must be carried out immediately.

Inspection and reporting

The outer ring track is examined for rust or etch spots and evaluated as shown in Table 4.3.

Note 3. To assist in estimating percentage corrosion area a transparent grid divided into suitable squares can be used.

Table 4.3. Rust or etch spot evaluation

Rating	Degree of rusting
(0)	No corrosion.
(1)	Not more than three small spots, each just sufficient to be visible to the naked eye.
(2)	Small areas of corrosion covering up to 1 per cent of the surface.
(3)	Areas of corrosion covering more than 1–5 per cent of the surface.
(4)	Areas of corrosion covering more than 5–10 per cent of the surface.
(5)	Areas of corrosion covering more than 10 per cent of the surface.

The various ratings are illustrated in Fig. 4.2. In all cases the area referred to represents the whole of the track.

Reporting

If the two ratings agree within the precision given for repeatability, report the two figures as the Dynamic Anti-Rust Test IP 220, otherwise make two further determinations.

Precision

Table 4.4 shows, for pairs of integer ratings, whether or not results should be considered suspect.

Note 4. The precision values have been obtained by statistical examination of inter-laboratory test results and were first published in 1967.

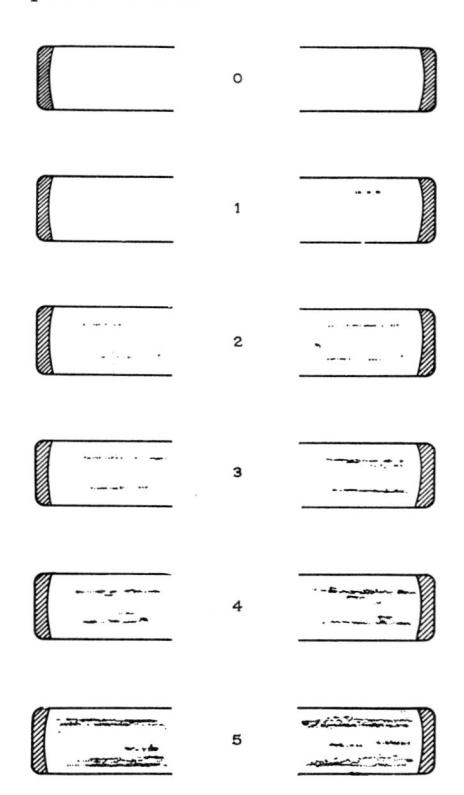

Fig. 4.2. Various ratings, illustrating degree of corrosion

Table 4.4. Suspect assessments for pairs of integer ratings

Pairs of ratings	Repeatability	Reproducibility
0 0	Not suspect	Not suspect
0 1	Suspect	Not suspect
0 2	Suspect	Suspect
0 3	Suspect	Suspect
0 4	Suspect	Suspect
0 5	Suspect	Suspect
1 1	Not suspect	Not suspect
1 2	Not suspect	Not suspect
1 3	Suspect	Not suspect
1 4	Suspect	Suspect
1 5	Suspect	Suspect
2 2	Not suspect	Not suspect
2 3	Not suspect	Not suspect
2 4	Suspect	Not suspect
2 5	Suspect	Suspect
3 3	Not suspect	Not suspect
3 4	Not suspect	Not suspect
3 5	Suspect	Suspect
4 4	Not suspect	Not suspect
4 5	Suspect	Not suspect
5 5	Not suspect	Not suspect

ADDITIONAL SPECIFICATIONS

Dynamic anti-rust test rigs

Fig. 4.3 shows part of a standard rig (Drawing No. SKF 1516600) which has eight self-aligning ball bearings (SKF 1306 K/236 725) fitted with adaptor sleeves and sleeve nuts (3) of polyamide plastic. The alternative rig has adaptor sleeves and nuts in metal. The bearings (2) are located in plummer blocks (1) (SKF SN507), which in the standard rig are polyamide plastic, according to drawing number SKF 720315. The alternative rig is completely metal. The rig is mounted on a machined steel plate (6), approximately $275 \times 985 \times 10$ mm. The shaft (4), 25 mm in diameter, is either coated with polyamide plastic (standard rig) or made of stainless steel (alternative rig). Two Vee-ring seals (5) per plummer block are required, together with a tool (7) for correct positioning of the seals.

Electric motor

Any suitable type may be used. The motor is fitted with reduction gear and flexible coupling to drive shaft directly at 80 rev/min.

Miscellaneous

A stand to hold the shaft on the work bench is also required and an automatic timing device is recommended.

APPENDIX 4.2
IP PRECISION EVALUATION PANEL

Dynamic anti-rust test

Thirty laboratories from several countries each tested eight greases in duplicate using rigs which contained either eight or four stations. On an eight-station rig, all greases were tested simultaneously, the assignment of greases to stations being randomized, and then the run was repeated, though with a different random assignment of greases to stations. On a four-station rig, four greases were tested simultaneously, and then tested again before the runs were repeated with the other four greases. Again, different random assignments were used for duplicate runs.

Results were ratings on the scale 0 to 5; 0 indicating a good grease and 5 a bad one.

There were two rig types, one with all parts made of steel, and the other with all parts made of nylon, except for the test bearing, which was steel. As well as calculating precision estimates, we were also asked to find if there was any difference between nylon and steel rigs. Laboratories 8 and 31 had both types of rig and provided results for all samples with each of them. Thus there were 32 pairs of results for each sample, these being tabulated in Table 4.2.

Analysis

The precision of a reported result x was proportional to

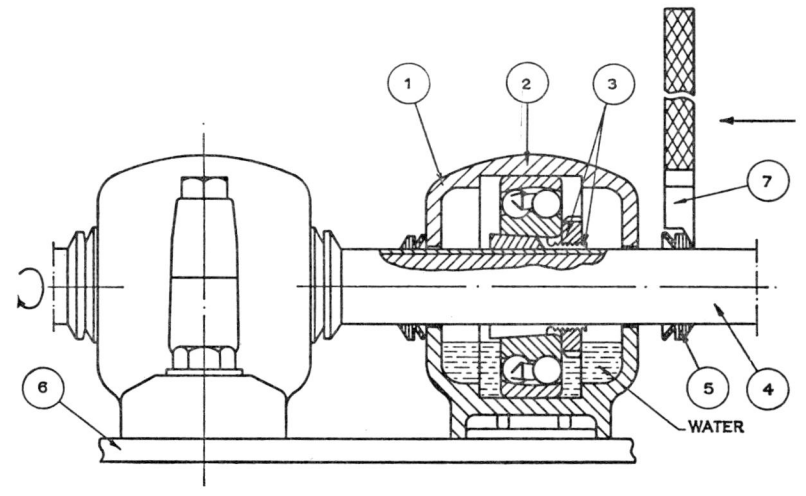

Fig. 4.3. Part of standard rig

$\sqrt{[x(5-x)]}$ and so to stabilize the variances the results were transformed to y's where

$$y = \arcsin\sqrt{\frac{x}{5}}$$

There were no outliers.

The analysis of variance is given in Table 4.5, and the laboratory means, in transformed data, in Table 4.6.

Discussion

There were significant differences between laboratories, though these differences varied from sample to sample. The variation between laboratories seemed to be random rather than due to any special factors such as location of laboratory. There was some evidence that steel rigs gave higher results than nylon rigs but this was not significant (mean nylon rating = 1·38, mean steel rating = 1·62).

These effects have all been taken into account when deriving the following precision estimates:

Repeatability = $0·334\{\sqrt{[x_1(5-x_2)]} + \sqrt{[x_2(5-x_1)]}\}$

with 256 degrees of freedom,

Reproducibility = $0·555\{\sqrt{[x_1(5-x_2)]} + \sqrt{[x_2(5-x_1)]}\}$

with approximately 330 degrees of freedom, where x_1 and x_2 are the pair of reported results.

(The reproducibility derived from the random effects model is

$$0·537\{\sqrt{[x_1(5-x_2)]} + \sqrt{[x_2(5-x_1)]}\}$$

with approximately 350 degrees of freedom.)

Table 4.5. Analysis of variance

Source of variation	D.F.	Sum of squares	Mean square	Expectation of mean square
Samples	7	470 284·250	—	
Laboratories	31	8 463·980	273·032	$\sigma^2 + 16\sigma_2^2$
Nylon vs. steel	1	778·184	778·184	
Rest	30	7 685·796	256·193	
Samples × laboratories	217	47 012·207	216·646	$\sigma^2 + 2\sigma_1^2$
Samples × nylon vs. steel	7	1 285·012	183·573	
Samples × rest	210	45 727·195	217·749	
Duplicates	256	12 557·309	49·052	σ^2
Total	511	538 317·773		

Table 4.6. Laboratory means (in transformed data)

Lab.	Rig type*	Country†	Mean	Lab.	Rig type*	Country†	Mean	Lab.	Rig type*	Country†	Mean
2	N	UK	43·5	12	U	H	29·6	23	S	F	41·0
3	N	UK	24·7	13	N	S	30·4	24	U	G	32·8
4	S	UK	33·8	14	N	S	30·4	25	U	G	34·4
5	S	UK	32·9	15	N	S	32·2	26	U	G	30·7
6	S	UK	31·2	16	N	S	36·3	27	U	G	37·9
7	S	UK	38·7	17	N	S	33·8	28	U	G	30·4
8	S	UK	32·1	18	N	S	33·0	29	U	G	34·5
8	N	UK	30·4	19	N	S	31·9	30	N	UK	36·3
9	N	UK	30·7	20	N	S	29·7	31	N	F	32·9
10	N	UK	23·9	21	N	S	25·5	31	S	F	33·4
11	N	H	32·1	22	S	F	33·7				

* N = nylon; S = steel; U = unknown.
† UK = United Kingdom; H = Holland; S = Sweden; F = France; G = Germany.

Table 4.7. Repeatability values

x_2			x_1			
	0	1	2	3	4	5
0	0	0·7	1·0	1·2	1·4	1·6
1	0·7	1·3	1·5	1·6	1·6	1·4
2	1·0	1·5	1·6	1·6	1·6	1·2
3	1·2	1·6	1·6	1·6	1·5	1·0
4	1·4	1·6	1·6	1·5	1·3	0·7
5	1·6	1·4	1·2	1·0	0·7	0

Table 4.8. Reproducibility values

x_2			x_1			
	0	1	2	3	4	5
0	0	1·2	1·7	2·1	2·4	2·7
1	1·2	2·2	2·5	2·7	2·7	2·4
2	1·7	2·5	2·7	2·7	2·7	2·1
3	2·1	2·7	2·7	2·7	2·5	1·7
4	2·4	2·7	2·7	2·5	2·2	1·2
5	2·7	2·4	2·1	1·7	1·2	0

Repeatability values for pairs of integer values of x_1 and x_2 are given in Table 4.7.

Thus, results obtained under repeatability conditions are suspect if they differ by more than 1, and also if they are 0 and 1 or 4 and 5.

Reproducibility values for pairs of integer values of x_1 and x_2 are given in Table 4.8.

Thus, single results from each of two laboratories are not suspect unless they differ by more than 2, or unless they are 0 and 2 or 3 and 5.

Conclusions

The following precision section is recommended for the method.

Precision: Duplicate results obtained under repeatability conditions (see Note 2 and Appendix E of the IP Standards) should be considered suspect if they differ by more than the following values:

Rating range	*Repeatability*
0–5	$0{\cdot}334\{\sqrt{[x_1(5-x_2)]}+\sqrt{[x_2(5-x_1)]}\}$

Single results submitted by each of two laboratories should not be considered suspect unless they differ by more than the following reproducibility values:

Rating value	*Reproducibility*
0–5	$0{\cdot}555\{\sqrt{[x_1(5-x_2)]}+\sqrt{[x_2(5-x_1)]}\}$

Note. These precision values have been obtained by statistical examination of inter-laboratory test results.

Table 4.4 shows, for pairs of integer ratings, whether or not results should be considered suspect.

APPENDIX 4.3

REFERENCE

(1) MENZIES, I. A. *Corrosion and protection of metals* 1965 (Iliffe Books Limited, London).

Paper 5

APPARENT (DYNAMIC) VISCOSITY AND YIELD STRENGTH OF GREASES AFTER PROLONGED SHEARING AT HIGH SHEAR RATES

G. J. Scholten*

Where sealed, grease-lubricated sleeve and ball-bearings are used with once-only lubrication for life, the rheological properties of the grease as a function of the service life are important in connection with the load-carrying capacity and the sealing capacity. To calculate the bearing capacity of a journal or spiral-groove bearing it is important to know the relation between the rate of shear and the shear stress in the lubricating film, i.e. the apparent viscosity, as a function of time. For sealing of the bearing the value of the yield strength of the grease, after shearing, is also important as a function of time. To measure these two magnitudes, a rotating-cylinder type of viscometer and a cylinder type of yield strength meter have been constructed. The viscometer permits the determination of viscosity up to shear rates of about 5×10^5 s^{-1}, while the yield strength meter can ascertain the ultimate strength, after stressing, with shear rates of up to 10^5 s^{-1}—both being determined within the temperature range 25–125°C. A number of commercially available greases have been tested.

INTRODUCTION

WHERE SMALL SHAFT DIAMETERS of 25 mm and less are concerned, there is an ever-increasing demand for silent, cheaper, hydrodynamic bearings which have a long service life (e.g. 10 000 h) and are lubricated for life with just some tens of cubic millimetres of grease.

The use of an appropriate lubricating grease in a hydrodynamic bearing has one main advantage over oil lubrication: the sealing is improved during rotation or standstill of shaft or bearing bush. Sealing in this context means the mechanism which either prevents leakage or reduces it to such an extent that a service life of, for example, 10 000 h or more can be expected.

When calculating the load-bearing and sealing capacity of this type of bearing it is important to know the dynamic viscosity and the yield strength of the lubricating grease, but these two magnitudes, as a function of temperature, shear rate, and duration of shear, are seldom indicated by the manufacturer. An idea can be obtained from the value of the dynamic viscosity of the base oil (the apparent viscosity of the grease is, in any case, greater) and the worked penetration after 100 000 double strokes in the A.S.T.M. 'grease worker' (a low penetration indicating a high yield strength). To allow the viscosity and yield strength to be measured during and after operation in conditions resembling, as far as possible, those prevailing in the grease-lubricated hydrodynamically operating bearing (1)–(4)† a rotating type of cylinder viscometer and a rotating type of cylinder yield strength meter were constructed.

The shear stability of a certain grease was judged by the change of the yield strength as a function of the shear time.

DESIGN CONSIDERATIONS AND CONSTRUCTION

Viscometer

As shown schematically in Fig. 5.1, the Couette viscometer basically consisted of a hollow shaft and a bearing bush. The shaft was connected by a Cardan-joint membrane coupling (not shown in the figure) to a d.c. shunt motor which had a tachogenerator. A heat shield was mounted between the shaft and the coupling. The bearing bush was provided with a heater element and holes for thermocouples, and was connected to a torque meter by

---

The MS. of this paper was received at the Institution on 15th September 1969 and accepted for publication on 14th October 1969.

* Research Engineer, N.V. Philips Gloeilampenfabrieken, Philips Research Laboratories, Eindhoven, Netherlands.

† References are given in Appendix 5.1.

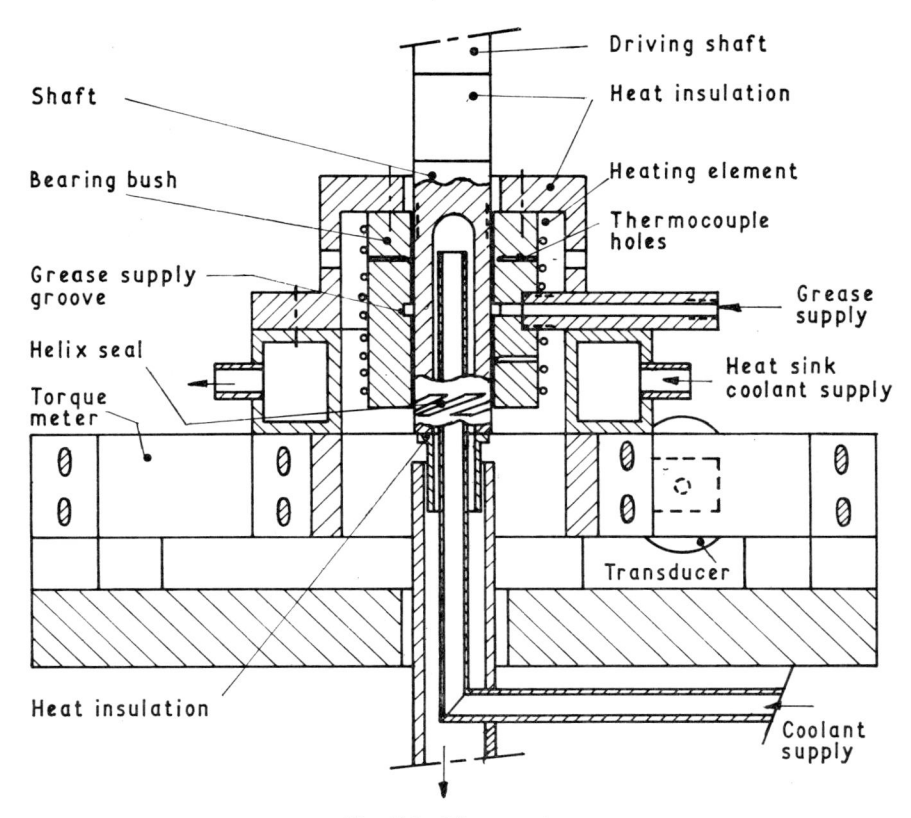

Fig. 5.1. Viscometer

way of a heat insulating bush and a heat sink. The insulations ensured a minimum axial flow of heat in the shaft and bush. The torque meter essentially consisted of three leaf springs and an inductive displacement pick-up.

The heat sink between the torque meter and the leaf springs quickly dissipated the heat that might cause the temperature of the leaf springs to rise. By this method, undesirable distortions of the springs, and therefore incorrect values of the moment measured, were prevented.

The apparent viscosity was determined by measurement of the moment, M, exerted on the bearing bush as a result of the viscous forces in the lubricating film when the shaft rotated at a speed n.

The apparent viscosity, η (N s/m^2) was determined from

$$\eta = C_1 \times \frac{M}{n} \qquad . \quad . \quad . \quad (5.1)$$

where C_1 is a constant of the instrument, M the measured moment (N m) and n the measured speed (rev/min). The shear stress, τ (N/m^2), was determined from

$$\tau = C_2 \times M \qquad . \quad . \quad . \quad (5.2)$$

where C_2 is a constant of the instrument.

Temperature effects

A number of temperature effects must be taken into consideration:

A temperature difference, ΔT_1, in consequence of a temperature gradient in the lubricating film

Because of the low coefficients of heat conduction of oil and grease, account should be taken of a temperature gradient in the lubricating film. This might give rise to an appreciable difference between the wall temperature at the film gap and the mean film temperature. On the assumption that there is a uniform wall temperature and a narrow gap between shaft and bushing, the maximum average increase in the temperature, ΔT_1 (degC), of the lubricating film (5) can be derived from

$$\Delta T_1 \approx 0.7 \times \frac{\eta D^2 \Delta r^2}{8\lambda} \qquad . \quad . \quad (5.3)$$

where η is the dynamic viscosity (N s/m^2), D the rate of shear (s^{-1}), Δr the radial gap between shaft and bush (m), and λ the heat conduction coefficient of the lubricant (J/m degC s).

An almost uniform wall temperature of shaft and bush was achieved by the choice of the same ratio of inner and outer diameters of shaft and bush, so that at an equal temperature of the outer side of the bush and the inner side of the shaft, the amount of heat dissipated radially was the same. The shaft and bush were made of the same material. Given a gap (Δr) of 11.6 μm, a rate of shear (D) of 7.2 × 10^4 s^{-1}, and a coefficient of heat conduction (λ) of 0.13 J/m degC s, a change in temperature (ΔT_1) of some 0·3 degC was to be expected with an SAE-40 oil at 25°C. Accordingly, the viscosity measured was about 1·5 per cent too low. This was no longer quite acceptable, since in equation (5.3) an isoviscous lubricant, i.e. a lubricant with constant viscosity, was considered. The error, however, was of a second order of magnitude.

A temperature difference, ΔT_2, resulting from an axial flow of heat

The bearing bush and the shaft, as already mentioned, were insulated axially to keep the axial heat flow to a minimum. The temperature measured at the sites of the lowest and uppermost thermocouple holes exhibited a maximum difference of 0·1 degC at a temperature of 75°C. Since the uppermost thermocouple hole was used in the measurements, the measured temperature was on average 0·05 degC too low. At 75°C the error in the measurement of η with an SAE-40 oil was only about 0·2 per cent, and this could be neglected.

Changes in the size of the gap of the bearing as a result of thermal expansion

The heat generated in the film gap had to be dissipated radially in the shaft and its bush. Consequently, a temperature difference was brought about between the inner and outer walls of the bush and the outer and inner walls of the (hollow) shaft. This difference in temperature led to thermal stresses, and the bushing and shaft did not expand equally.

Let the following conditions apply:

(a) a temperature difference, ΔT_3, in shaft and bush;
(b) bush and shaft in thermal equilibrium;
(c) an identical ratio of inner and outer radii of shaft and bush; and
(d) both shaft and bush made from the same material.

With the above assumptions the following relation (**6**) will hold to a good degree of approximation (in m):

$$\Delta(\Delta r) = -0.2 \times \alpha \times \Delta T_3 \times r \quad . \quad (5.4)$$

where $\Delta(\Delta r)$ is the change in the film gap, α the coefficient of linear expansion (m/m degC), and r the radius at the site of the film gap (metres).

For $\Delta T_3 = 1.5$ degC—the maximum temperature difference measured—the reduction of the gap in steel would be about 0·03 μm. This is negligibly small in comparison with the gap (Δr) of 11·6 μm used here.

A temperature difference, ΔT_4, resulting from a delay in temperature measurement

The viscosity was measured, particularly at the lower temperatures up to 50°C, during heating up of shaft and bush. This means that for a measuring temperature of, say, 25°C the motor was suddenly started at a shaft and bush temperature of, say, 24°C. Immediately the desired temperature of 25°C was reached, the motor was stopped and measurement took place. This had the advantage of simplicity of the measuring equipment. It was then necessary, however, to take into account a time delay in the measurement of the temperature, and the real temperature of the wall, and thus of the lubricant, was consequently ΔT_4 degC higher than the measured temperature. The measured values of viscosity and shear stress would then be lower than the actual values. With the help of a Newtonian oil it was possible to determine experimentally the magnitude of this departure. There is general agreement that in the case of an ordinary lubricating oil no viscosity drop occurs below Bondi's critical shearing stress of 5×10^4 N/m² (**7**). This could, however, be checked theoretically as follows.

A thin layer of shaft and bush, measured from the film gap, could be considered as flat and treated as a semi-infinite solid. With a semi-infinite solid the rise in the surface temperature, ΔT_4, could be calculated after a sudden and constant supply of heat, distributed uniformly over the surface. The following relation (**8**) would then apply:

$$\Delta T_4 = \frac{1.13 \times Q}{\lambda}(a \times t_q)^{1/2} \quad . \quad . \quad (5.5)$$

where Q is the heat flux (J/m² s), λ the coefficient of thermal conductance (J/m degC s), a the thermal diffusivity coefficient (m²/s), and t_q the delay time as a function of the heat flux (s).

Over a temperature range of, for example, 1 degC the power developed in an SAE-40 oil at 25°C changed by a mere 5 per cent, so that Q could, without excessive error, be considered constant. By allowing the driving motor to reach its operating speed suddenly, which was in practice achieved within 0·1 s, measurement was made of the time that elapsed before the change in temperature was detected. This delay time appeared to depend on the power generated.

For example, with 30 W/m² the delay measured was $t = 0.8$–0·9 s. At the instant that a rise in temperature was detected, the temperature of the walls of the shaft and the bush was ΔT_4 degC higher at the site of the lubricating film and could thus be calculated by means of equation (5.5). If ΔT_1 and ΔT_4 are taken together, the percentage error in the measured viscosity and the measured shear stress can be determined. An ordinary lubricating oil was considered to be Newtonian, since the measured change, within a measuring accuracy of ±2 per cent, was in agreement with the calculated change.

Fig. 5.2 is a plot of the percentage correction on the measured values of τ and η as a function of the product of η and D^2, which is proportional to the generated power [see equations (5.3) and (5.5)].

The grease supply groove

The bearing bush was provided halfway up with a circular grease supply groove (Fig. 5.1). This was necessary, when filling with grease, to permit the air in the gap to be axially displaced. The supply hole was closed after the film gap had been filled. For the grease in this groove the rate of shear, D, could fall to as little as 150 s⁻¹. In the case

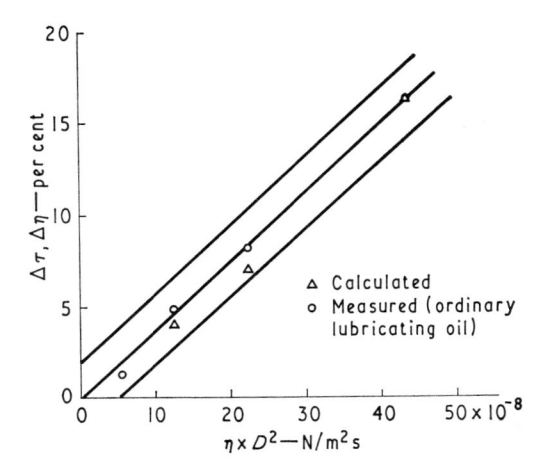

Fig. 5.2. Correction graph

of a pseudo-plastic grease this was accompanied by a higher viscosity with a relatively high shear stress. Measurements at shear rates of less than 3×10^3 s^{-1} were, however, no longer considered reliable. By linear extrapolation, to $D = 0$, of the measured relation between τ and D, and by assuming that the value of τ thus found was the maximum possible value in the grease groove, an upper limit of the error resulting from this groove could be determined.

Beyond a rate of shear, D, of about 3×10^4 s^{-1} this error was found to be only exceptionally greater than 1 per cent, and as a rule it was less than 1 per cent. Up to a value of $D = 3 \times 10^3$ s^{-1} this error limit could rise to about 4 per cent.

Yield strength meter

The yield strength meter, too, essentially consisted of a shaft and a bearing bush (Fig. 5.3). During shearing, the

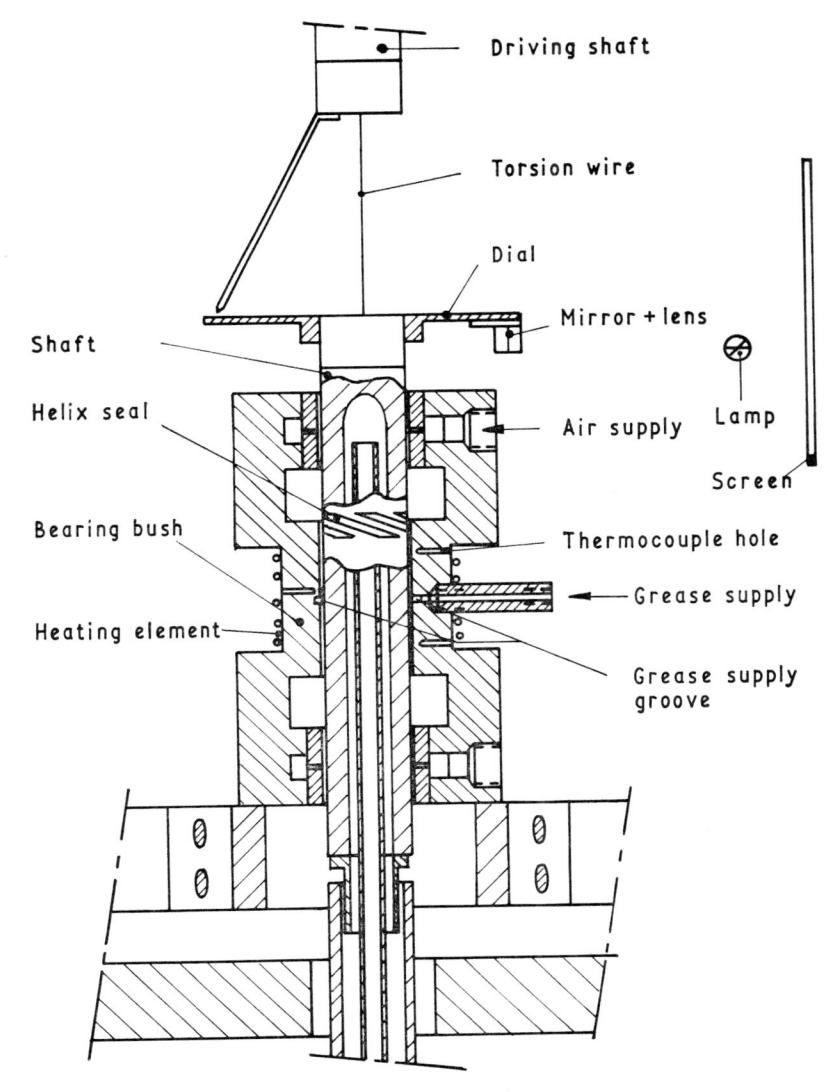

Fig. 5.3. Yield strength meter

shaft was driven in exactly the same manner as for the measurement of the viscosity in the viscometer.

Measurement of the yield strength after shearing now took place with a torsion-wire meter that was mounted between the flexible coupling and the shaft (Fig. 5.2).

The bearing bush consisted of three bearings in line. The two outer were aerostatic bearings, and the one in the centre was the measuring bearing proper, provided once more with thermocouple holes, a heater element, and a grease supply groove. The purpose of the air bearings was to ensure that during measurement of the yield strength (with the shaft not rotating) the shaft did not touch the bearing bush—which, in the viscometer, was doubtful.

Care had to be taken to prevent axial displacement of the driven shaft during connection to and disconnection from the driving motor of the torsion-wire meter. The torsion-wire meter consisted in principle of a torsion-wire, a pointer, a dial (a disc with a scale graduated in degrees), and a lens-and-mirror system. By means of the lens-and-mirror system (fixed to the driven shaft), a lamp, and a projection screen it was possible to detect an incipient rotation of 10 seconds of arc of the driven shaft. For performing the measurement, the rotor of the driving motor was turned slowly by hand until movement of the driven shaft was detected. The torsion angle was read off with the aid of the dial and pointer. The rotor was then slowly turned in the opposite direction, and the torsion angle was read off once more. The average of these clockwise and anticlockwise rotations was a measure of the torque applied to the driven shaft. The average angle of rotation at which the shaft began to slip—ascertained after about 10 seconds of arc at least, corresponding to a shear strain of about 10 per cent—was used as a measure of the yield strength (N/m^2):

$$\tau_y = C_3 \times \gamma \quad . \quad . \quad . \quad . \quad (5.6)$$

where C_3 is a constant of the instrument and γ is the average torsion angle.

Effect of the grease supply groove

The grease supply groove had also to be considered in measurements of the yield strength. This groove accounted for 3·5 per cent of the overall length of the measuring bearing. On the assumption that the grease in this groove was not broken down (on account of the low rate of shear) it was necessary in principle to subtract 3·5 per cent of the original unsheared value of τ_y from the measured value after pre-shearing. If the original value was 200 N/m^2, for example, and the measured value after shearing was 10 N/m^2, then the actual yield strength would be $10 - 3·5 \times 200/100 = 3$ N/m^2. Grease, however, is always more or less elastic, and the effect of this elasticity was measured with another shaft, which at the site of the measuring bearing had such a diameter that the gap was 1 mm.

This gap equals the gap opposite the grease supply groove when use is made of the normal measuring shaft. It was found that the force now required to yield a rotation of 10 seconds of arc, corresponding to a shear strain of about 0·1 per cent, was at the most 5 per cent of the force needed to reach the ultimate yield strength. The above error of 3·5 per cent was therefore reduced to $\frac{1}{20} \times 3·5$, i.e. about 0·2 per cent. The maximum corrected value in the above example would therefore be 0·4 N/m^2 lower than the observed value of 10 N/m^2. Hence the above assumption was exaggerated.

Given a reading error of the angle of rotation of $\pm 1°$, the possible measuring error would be between ± 20 per cent at the lower values of the yield strength (to about 3 N/m^2) and ± 2 per cent at the higher values (to about 200 N/m^2).

Temperature effect

A temperature effect might influence the measurement. The axial dissipation of heat through the two aerostatic bearings (Fig. 5.3) would give rise to a temperature difference between the outer ends of the measuring bearing and its centre. This difference was 3 degC at a temperature of 75°C. The mean temperature error was then about +1·5 degC. At 50°C the difference was only +0·5 degC, so that the mean error was 0·25 degC. Because the temperature was measured on the uppermost thermocouple hole, placed at 1·5 mm above the middle of the upper half of the bearing (see Fig. 5.2), the measuring temperatures were taken, respectively, as 74·8 and 49·95°C (in practice 50°C).

GENERAL NOTES

The bearing bushes and shafts were made of nitrided steel and lapped to a surface roughness of 0·5–1 μin c.l.a. They were straight and round to within a few tenths of a micron.

The shafts were provided at the ends of the operating parts with 1·5-mm long helix seals, put on by means of photochemical etching (see Figs 5.1 and 5.3). These seals were necessary to prevent leakage during the time of shearing. The helix seals, however, provided about 20 per cent less friction than an ungrooved shaft over the same length; accordingly, the effective length of the bearing bush had to be reduced by 0·6 mm. On account of the pumping pressure generated by the helix seals during rotation up to 2×10^5 N/m^2, the grease supply groove had to be closed after the film gap had been filled.

By means of a thermostat (consisting of an NTC resistor at a few tenths of a millimetre from the lubricating film, an amplifier, and an electromagnetically operated water valve) the temperature at which the grease was sheared was kept to within 1 or 2 degC.

Filling of the gap was checked by means of a transparent model of the bearing bush. During filling it was necessary to allow the shaft to rotate slowly, otherwise the grease might not completely fill the gap, which could result in a certain amount of air being entrapped.

To check the effect of the helix seals, ordinary lubricating oil was subjected to shearing for several hours. No apparent drop in viscosity, which might occur in consequence of leakage, was detected. Leakage during standstill was prevented by capillary forces and by the yield strength, if present after shearing of the grease.

For each new measuring temperature fresh grease was put into the gap for the following reason. On account of the rather large coefficient of thermal expansion of the grease, grease which had received little pre-shearing would be pressed out of the grease supply groove into the gap when the temperature is raised from, for example, 25°C to 50°C, or the gap might partly be drained when the temperature is lowered. In the former case the grease would have had an unknown history so that comparative measurement would be impossible, and in the latter case too low a value of η and τ would be found.

Both in the viscometer and in the yield strength meter the capacitance of the capacitor formed by the shaft and bush was measured continuously during shearing and measuring. A measurable value of the capacitance guaranteed hydrodynamic lubrication in measurements with the viscometer and no fouling of the shaft in the three bearings of the yield strength meter.

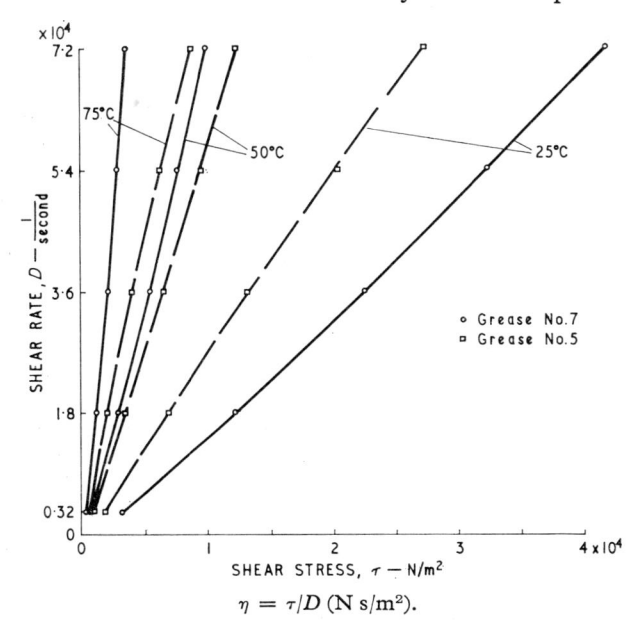

Fig. 5.4. Shear data

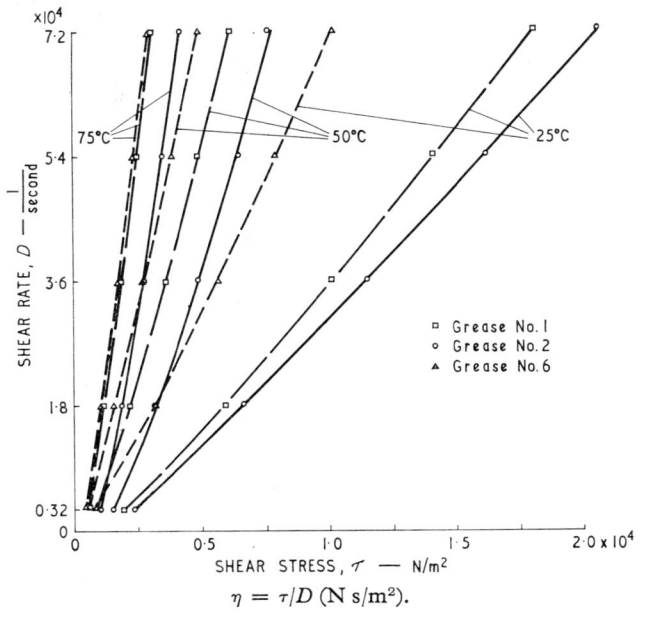

Fig. 5.5. Shear data

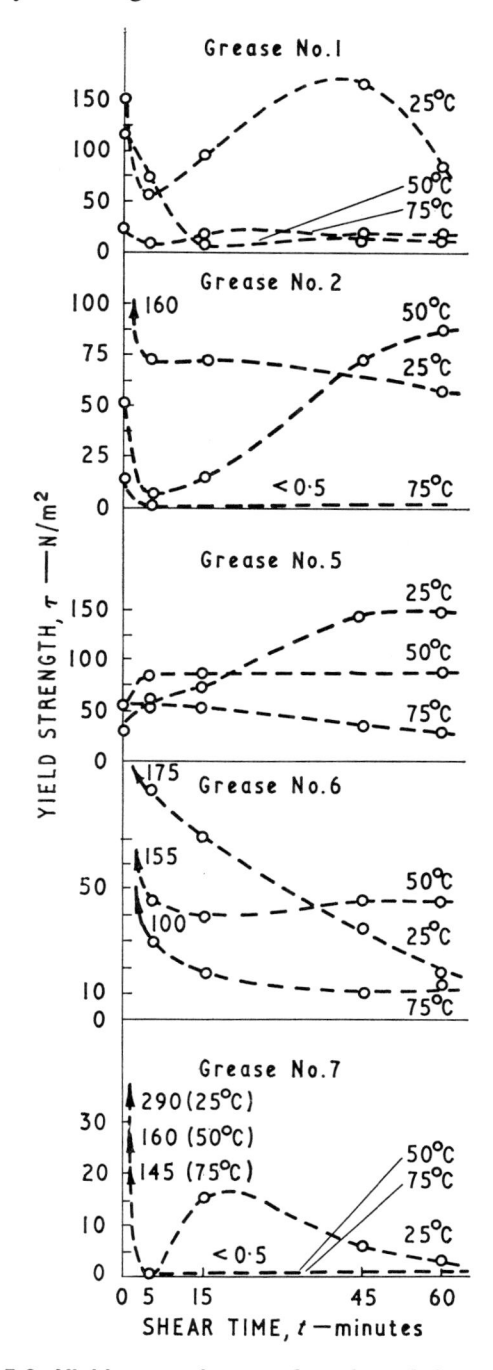

Fig. 5.6. Yield strength τ as a function of shear time t

TEST PROCEDURE

Yield strength

Each of the investigated greases was sheared at 25, 50, and 75°C (± 1 degC) with a shearing rate $D = 7.2 \times 10^4$ s^{-1}. After shearing times of respectively 0, 5, 15, 45, and 60 min the yield strength was measured at the three above temperatures within 5 min of stopping of the driving motor. Five minutes after the last measurement at a certain temperature the last measurement was repeated, in order to ascertain whether there was recovery of the yield strength.

Shear stress and viscosity

Each of the investigated greases was sheared, just as for determining the yield strength, until at a shear rate $D = 7.2 \times 10^4$ s^{-1} no change could be determined any longer in the measured moment. Then at the temperature of shearing (now kept constant within 0.15 degC) the moment was measured at shearing rates of 7.2×10^4, 5.4×10^4, 3.6×10^4, 1.8×10^4, and 0.32×10^4 s^{-1}.

RESULTS

The results in respect of the shear stress (τ) as a function of shear rate (D) are given in Figs 5.4 and 5.5. The apparent viscosity can be determined by means of the relation:

$$\eta = \frac{\tau}{D} \quad \quad \quad (5.7)$$

The effect of the shear time (t) on the yield strength (τ_y) is given in Fig. 5.6.

DISCUSSION

Li-soap and Ca-soap greases showed considerable shear thickening within a shear period of 1 hour. Furthermore, the generated torque became so irregular in magnitude that there was little point in performing the measurements (see Table 5.1, grease No. 3 and No. 4). These kinds of grease would seem unsuitable for use in hydrodynamic bearings having a narrow lubricating gap.

The measurements of the yield strength revealed that on most occasions the yield strength showed a rapid drop during the first moments of shearing. The drop in the yield strength was accompanied by a marked drop in the shear stress during the first moments of shearing. Bauer, Finkelstein and Wiberley [9] also found a great decrease in the apparent viscosity in the first minutes of shearing. They utilized a cone-and-plate viscometer which allowed shear rates up to 2×10^4 s^{-1}. The yield strength need not always decrease continuously with an increasing shear time; after an initial rapid fall a temporary rise in the yield strength was sometimes observed. Moore and Cravath [10] found such a change in the penetration as a function of the time of stressing. They utilized a Shell roll tester in which some 75 g grease were stressed in shear. This amount, however, was not stressed continuously in their appliance, in contrast to the viscometer described here, so that they found appreciably longer shear times. The quantity subjected to shear in the present viscometer was some 20 mg. The yield strength is in part determined by the temperature at which shearing takes place (see Fig. 5.6, grease No. 2). At 25 and 75°C continuous breakdown occurred during

Table 5.1

Grease No.	Type of thickener	Base oil, type and content, per cent	Unworked penetration	Base oil viscosity, N s/m²	Apparent viscosity at a shear rate of 7.2×10^4 s^{-1}, N s/m²
1	Na-complex soap	Mineral . . . 86	260	25°C 0.095 50°C 0.029 75°C 0.013	0.253 0.087 0.044
2	Ba-complex soap	Mineral . . . 77	260	25°C 0.095 50°C 0.029 75°C 0.013	0.286 0.107 0.059
3	Li-soap	Mineral synthetic . 85	280	25°C 0.068 50°C 0.019 75°C 0.008	— — —
4	Ca-soap	Mineral . . . 87	280	25°C 0.090 50°C 0.024 75°C 0.009^5	— — —
5	Gel of bentonite	Synthetic . . 90	400	25°C 0.048 50°C 0.019 75°C 0.008^5	0.383 0.172 0.122
6	Clay	Synthetic	300	25°C 0.045 50°C 0.017 75°C 0.008^5	0.143 0.069^5 0.042
7	Silica	Mineral . . . 95	390	25°C 0.290 50°C 0.056 75°C 0.018	0.585 0.140 0.51^5

1 hour of shearing, whilst at 50°C a rise in the yield strengths was observed after the initial period of degradation. All the greases investigated behaved more or less pseudo-plastically. An exception was provided by grease No. 5 which, at 25 and 75°C, exhibited slight shear thickening (see Figs 5.4 and 5.5).

The apparent viscosity of the grease can be considerably higher than the dynamic viscosity of the base oil. In Table 5.1 the apparent viscosity is given, where possible, for a shear rate of 7.2×10^4 s^{-1}, in addition to the dynamic viscosity of the base oil. In many cases the apparent viscosity of soap base greases at shear rates above 10^4 s^{-1} is about 2·5 to 4 times the viscosity of the base oil.

Furthermore, there is no correlation between penetration and measured yield strength (Table 5.1 and Fig. 5.6) and no correlation between penetration and measured viscosity after pre-shearing (Table 5.1). A grease rated as good or rather good with respect to mechanical stability measured with standard equipment can earn a very low rating in the test rig described (for example, grease No. 7).

The relation between τ and D, and the value of τ_y, should be determined each time a new batch of a grease is received. The results of the measurements sometimes show discrepancies of a factor of 2 or more between different batches of the 'same' grease.

Though the average time of shear was only about 1 hour, that period gives an idea of the mechanical stability of greases that are to be used in hydrodynamic bearings, and this can serve as a basis for a preliminary selection.

APPENDIX 5.1

REFERENCES

(1) MUYDERMAN, E. A. 'Spiral groove bearings', thesis, Delft University of Technology, The Netherlands, 1964 (March); published in U.S.A. and Canada by Springer Verlag Inc., New York; published in Great Britain by Macmillan & Co. Ltd, London.

(2) REINHOUDT, J. P. 'A grease-lubricated hydrodynamic bearing system for a satellite flywheel' (to be published in *ASLE. Trans.*).

(3) MUYDERMAN, E. A. 'New possibilities for the solution of bearing problems by means of the spiral-groove principle', *Lubrication and Wear Fourth Conv., Proc. Instn mech. Engrs* 1965–66 **180** (Pt 3K), 174.

(4) HOROWITZ, H. H. and STEIDLER, F. E. 'Calculated performance of greases in journal bearings', *ASLE Trans.* 1963 **6**, 239.

(5) BARBER, E. M., MUENGER, J. R. and VILLFORTH, F. J., Jun. 'A high rate of shear rotation viscometer', *Analyt. Chem.* 1955 **27** (No. 3), 425.

(6) BOLEY, B. A. and WEINER, J. H. *Theory of thermal stresses* 1960 (Wiley & Sons, New York).

(7) BLOK, H. 'Viscosity of lubricating oils at high rates of shear', *Ingenieur, 's Grav., Materialenkennis* 1948 **5** (in English).

(8) CARSLAW, H. S. and JAEGER, J. C. *Conduction of heat in solids* 1959 (Clarendon Press, Oxford).

(9) BAUER, W. H., FINKELSTEIN, A. P. and WIBERLEY, S. E. 'Flow properties of lithium stearate–oil model greases as functions of soap concentration and temperature', *ASLE Trans.* 1960 **3**, 215.

(10) MOORE, R. J. and CRAVATH, A. M. 'Mechanical breakdown of soap base greases', *Ind. Engng Chem.* 1951 **43** (No. 12), 2892.

Paper 6

GREASES AS LUBRICANTS FOR METAL AND PLASTIC-LINED PLAIN BEARINGS

J. G. M. Sallis*   W. H. Wilson†

Compared with data on oil-lubricated plain bearing materials, very little has been published about the performance of conventional plain bearing materials with grease lubrication. This paper presents the results of wear tests on bushes as a guide for designers to the selection of suitable materials and types of grease. Marked differences were shown between the relative performance of different metals lubricated with grease compared with those lubricated with oil. Good mechanical and chemical stability are most important for greases used to lubricate plastic bearings. To stimulate constructive comment, a suggested ideal grease specification for plain bearings is mooted.

THE THEORY AND APPLICATION OF GREASE TO PLAIN BEARINGS

PLAIN BEARINGS are very widely operated with grease lubrication; yet they usually run with lubricants developed not for their own requirements but to suit the needs of rolling element bearings.

This impression of second-class bearing status is heightened by the remarkable neglect of grease-lubricated plain bearings by theoretical tribologists. They are dismissed in a few sentences in most treatises on plain bearings merely as being useful for low-speed high-load applications, for use in dirty environments, or where sealing arrangements prevent the use of oil. In the best known of the few published papers on grease-lubricated plain bearings, Cohn and Ohren report experimental investigations of a rather non-typical condition where a journal bearing is copiously applied with grease and is operating at low to moderate speeds with light loading (1)‡. This shows that hydrodynamic or rheodynamic lubrication (2) does occur but that the pressure distribution curve is flatter and more extensive than with oil, because of the reduced end leakage from the bearings (Fig. 6.1).

Simple hydrodynamic lubrication theory assumes the lubricant to behave like a Newtonian liquid, i.e. stress is proportional to rate of strain. In Fig. 6.2 this would be a straight line through the axis with its slope the coefficient of viscosity. Fig. 6.2 also shows that to assume grease behaves similarly to a Bingham plastic is a reasonable approximation.

Theoretical flow characteristics are represented by

$$\tau = \mu \frac{\partial u}{\partial y} \quad \text{(for an oil)} \quad . \quad . \quad (1)$$

$$\tau = \tau_0 + \eta \frac{\partial u}{\partial y} \quad \text{(for a grease)} \quad . \quad . \quad (2)$$

when $\partial u/\partial y \neq 0$

where τ is the shear stress in the lubricant, τ_0 the yield stress of a Bingham solid, μ the viscosity of a Newtonian fluid, η the plastic viscosity of a Bingham solid, u the velocity in direction X, and y the distance normal to the bearing surface.

Milne (2), Chakrabarti and Harker (3), Osterle et al. (4), and Slibar and Paslay (5) all treat grease as a Bingham plastic. The former shows that for a simple slider bearing a grease in the sheared regions operates in a similar manner to an oil with equivalent viscosity. The plastic viscosity of the grease is shown to govern the pressure generated, and the frictional drag is increased by an amount dependent on the yield value for the grease. Unsheared regions within the grease were predicted and shown to be the cause of the flatter pressure curves. These unsheared regions, or cores, were also observed from a transparent model used to check the predictions qualitatively.

The MS. of this paper was received at the Institution on 3rd September 1969 and accepted for publication on 11th November 1969. 33
* *Senior Research Investigator, Glacier Metal Company, Ealing Road, Alperton, Wembley, Middlesex.*
† *Industrial Unit of Tribology, Leeds; formerly Senior Materials Engineer, Glacier Metal Company, Ealing Road, Alperton, Wembley, Middlesex.*
‡ *References are given in Appendix 6.1.*

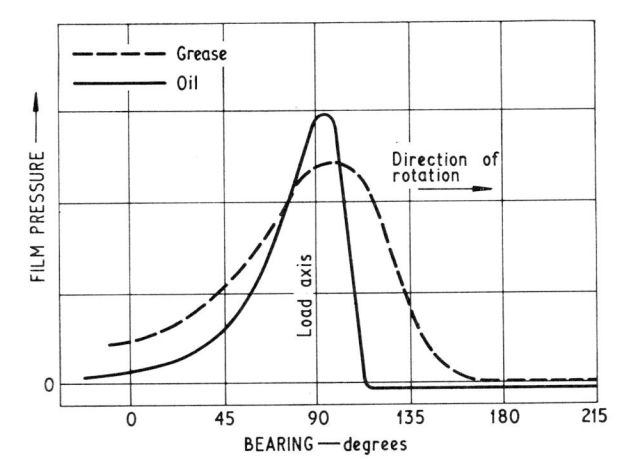

Clearance, 0·8 per cent.
Load, 40 lbf/in² (275 kN/m²).
Speed, 350 rev/min.
Nominal diameter, 1 in (25·4 mm).
$l/d = 1$.

Grease: NGLI 2 with 15 per cent Ca and Na soap and an SAE 20 mineral oil.
Oil: SAE 20.

Fig. 6.1. Comparison of the lubricant pressure distribution around a journal bearing, measured half-way along the bearing, for an oil and for a grease (from Cohn and Ohren (1))

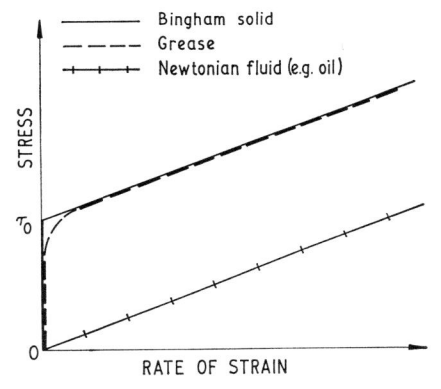

Fig. 6.2. Stress versus rate of strain relationship of a Bingham solid and a Newtonian fluid compared with the apparent curve for a grease measured when using a capillary-flow-type viscometer

References (4) and (5) present analyses of theoretical aspects of grease-lubricated thrust bearings.

Practical performance data of use to designers concerned with greases and grease-lubricated journal bearings are contained in papers by Sims and Hunt (6), Chakrabarti and Harker (3), and Bradford, Barber and Muenger (7).

Sims and Hunt record the results of tests on a grease-lubricated journal bearing under high cyclic load low-speed operation, and they show the importance of choosing the correct lubricant for the prevailing conditions. Friction coefficients less than 0·01 were possible with loads over 1000 lbf/in² (6·9 MN/m²) and surface speeds around 40 ft/min (0·20 m/s).

Bradford, Barber and Muenger's tests were run at higher surface speeds (up to 900 ft/min or 4·6 m/s) and lighter loads. They found the friction of a steadily loaded grease-film-lubricated journal bearing was affected by the shear history of the lubricant. In addition, it was determined that although there were similarities with oil-lubricated bearings, use of the familiar ZN/P parameter often did not reduce bearing test data to a single curve.

Chakrabarti and Harker (3) concluded from experimental results gained on a steady radially loaded grease-lubricated journal bearing test machine that above 100 ft/min (0·51 m/s) surface speed grease behaves as a Bingham plastic with friction torque increasing linearly with speed. They derived an equation for the friction torque with experimentally derived coefficients, in the form

$$Tf = Aw + BWw + CW + D$$

where Tf is the friction torque, w the journal angular velocity, W the radial load, A is proportional to the plastic viscosity, and D is proportional to τ_0, the yield stress.

A great deal of experimental and theoretical examination of plain bearings with grease lubrication remains to be carried out.

In practice, a grease-lubricated bearing is seldom full of grease. Thus, a theory of grease lubrication is needed which would enable designers to predict performance characteristics of bearings running under the more common marginal, boundary, or mixed lubrication conditions.

Typical applications of grease in plain bearings include those taking advantage of:

(*a*) its semi-solid structure to minimize leakage and throw-off, and hence to prevent contamination of the environment, or to give extended running life without relubrication;

(*b*) its sealing properties, to allow operation in dirty or wet surroundings or to reduce costs by simplifying designs;

(*c*) its ability to form a corrosion protective film;

(*d*) its role as a lubricant reservoir supplying fresh lubricant to the bearing as it is required; and

(*e*) the adherent film it forms over bearing surfaces giving good boundary lubrication, tolerance of shock loading, low friction at start-up, and a fluid film lubrication range extended into the low-speed high-load area. This characteristic of the performance of greases is suggestive of the squeeze film or elastohydrodynamic lubrication phenomena found with oil-lubricated bearings and would result, at least in part, from the marked increase in apparent viscosity found to occur with a grease at low rates of shear (6).

Greases are also used for economic reasons. They offer low rates of lubricant consumption and can operate effectively in relatively crude mechanisms where clearances may be large.

Some of the disadvantages of using grease arise directly from aspects of their otherwise desirable characteristics. They produce higher friction losses than oil (7), at the same time giving much poorer heat removal from the bearing, because there is little flow of lubricant over the bearing surface. Their mechanical stability is relatively poor, and they tend to deteriorate when operating at high velocities and high temperatures.

PERFORMANCE TESTING AND TEST MACHINES

The evaluation of greases as lubricants has been almost exclusively aimed at their performance when lubricating gears or rolling contact bearings, in which they have two distinct roles. First, as a reservoir of virtually unworked grease, part of which is slowly fed to the surface in contact; and second, as an oily thoroughly overworked grease forming the lubricating film. The ratio of unworked to worked grease is large.

With plain bearings the degree of working and the magnitude of the shear forces in the thin load-supporting lubricant film are greater, and the ratio of reserve to worked grease is low, being limited to that which is contained in the clearance space and in any grooving or other reservoirs in the breaking surface. Together with the higher film temperatures, these factors ensure that use in plain bearings is a very severe test of a grease.

The grease performance comparisons presented later were based on initially lubricated bush tests run under laboratory conditions on three types of specially designed machines called Libra, Beta, and Zircon.

Reliable assessment and comparison of the performance of bearing material/grease combinations involve thousands of hours of testing. The attractive economies and saving in time offered by accelerated testing, i.e. testing using loads and speeds sufficiently high to wear out the bearing in a much shorter time than would be acceptable in practice, must be resisted. In addition, the methods of test must be carefully chosen so that the wear mechanisms involved, the relative areas of mating surfaces, pressures, relative motions, etc., do not differ fundamentally from those likely to arise in practice. Measurement of friction, wear, and bearing temperature should be possible during the course of a test.

The basic layout of the Libra type of test machine is shown in Fig. 6.3. A cantilever test shaft fits in a spindle driven by a pulley and vee-belts. The $\frac{5}{8}$-in (16-mm) bore, $\frac{3}{4}$-in (19-mm) long test bush is housed in a self-aligning load yoke and bears against the En 1A test shaft. Journal surface finish is 6 μin c.l.a. (0·15 μm).

Loading is downwards by equal weights attached to load hangers on either side of the load yoke; friction torque is measured by the spring-balance tension needed to prevent rotation of the load yoke. Back of bush temperature on the line of load is sensed with a thermocouple and recorded continuously.

The Beta test machine differed in having the load

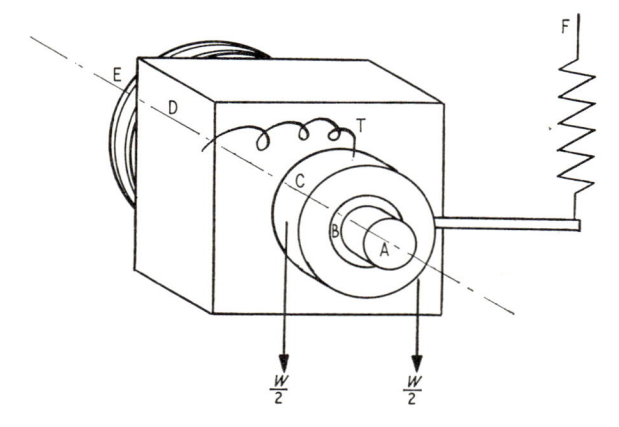

A Replaceable test shaft.
B Test bush.
C Load yoke.
D Support bearing housing.
E Drive pulley.
F Friction torque measurement.
T Back of bush thermocouple.
W Load.

Fig. 6.3. Essential features of a test head of the Libra machine

applied hydraulically upwards on a rigid test bush housing. Friction torque measurement was not possible.

Zircon test machines provide a quick, simple, and somewhat crude method for ranking bearing materials. The test consisted of loading a flat specimen clamped to a hinged lever against a rotating 1-in diameter silver steel shaft mounted in plummer blocks. Lubricant was applied at start-up only. Performance was judged by the width of the wear scar produced after a standard length of test.

GREASE-LUBRICATED PLASTIC BEARINGS

Plastic-based materials are now well established for use as plain bearings. Their good compatibility with metallic mating surfaces has led to the widespread use of some types as dry bearings, with no lubrication at all. The main limitation on performance is then set by the temperature of the bearings, caused by dry frictional heat generation. Even a very small amount of lubricant introduced between the mating surfaces will generally cause a substantial reduction in the friction coefficient of a plastic bearing, thereby greatly increasing its performance potential.

The role of a lubricant for plastic bearings is to establish, and maintain, an adherent friction-reducing film on the plastic surface. Except in steadily rotating bearings flooded with oil, this film is usually too thin to separate completely the two bearing surfaces. However, very low friction and wear can be achieved with plastic bearings under extreme conditions where metal on metal bearings would tend to scuff or seize (Table 6.1) (8).

Glacier DX in Table 6.1 is a proprietary polyacetal plastic-lined bearing material which exploits the ability of plastics to operate with minimal lubrication. It comprises a 0·014-in (0·36-mm) layer of plastic bonded to a steel

Table 6.1. Comparison of Glacier DX and common metal bearing materials tested on the Libra machine with initial grease lubrication only at 20 000 PV*, 130 ft/min (0·66 m/s), 63 μm diametral clearance

Bearing type	Material	Mean time to failure, h	No. of results	Steady running temperature, °C
Plain bore bush	DX	1900†	2	30
Plain bore bush	7½/3/89½ Tin-based Babbitt (SAE 12)	46	2	30
Plain bore bush	89½/10/½ Phosphor-bronze	69	2	38
Thrust washer	Grooved DX	800	6‡	65
Thrust washer	Oil-impregnated porous bronze	20	1	54
Thrust washer	Tin-based Babbitt	23	2	70
Thrust washer	Phosphor-bronze	22	2	100

* 1 PV has units lbf/in^2 × ft/min and is equivalent to 35 N/m^2 × m/s.
† Wear rate after bedding-in 0·3 × 10^{-3} in per 1000 h.
‡ Standard deviation (after transforming to logarithms), 0·95.

backing by means of a porous sintered bronze layer, thereby making a product which can be formed, machined, and housed in much the same manner as conventional bearing liners. Other high melting point thermoplastics, such as nylon or polyester, would give a similar bearing performance to polyacetal in the same form and under the same conditions.

The needs of a plastic bearing in the way of a lubricant appear to differ from the needs of metal bearings running under the same marginal lubrication conditions. Because of the inherent good bearing compatibility with steel of both thermosetting and thermoplastic polymers there is no need for chemically active anti-wear or extreme pressure additives; indeed, such additives could conceivably be detrimental and cause chemical breakdown of the polymer. In bearings running with initial lubrication only, or with infrequent relubrication, the grease requires to be exceptionally stable to withstand, without separation or over-softening, the very severe shearing which occurs inevitably in a plain bearing. This is greatly aggravated by both the extremely thin lubricating films (probably less than 1 μm thick) and the small total amount of lubricant present in the bearing. In addition, the lubricant must be inhibited against oxidation to remain effective over a period of several thousand hours' operation without replenishment (continuous running for 17 000 h has been achieved with a Glacier bush lubricated before start-up only). The reservoir of lubricant in the bearing clearance space, grooving, etc., must remain in close proximity to the loaded area, to be readily available when required throughout the life of the bearing. The importance of keeping the lubricant in the right place is clearly illustrated in Fig. 6.4. With upward loading the lubricant tends to drain under the effect of gravity towards the loaded area, giving lives up to a factor of 10 greater than with downward loading, where the drainage is away from the loaded area.

EFFECT OF GREASE TYPE ON THE PERFORMANCE OF PLASTIC BEARINGS

Tests with a wide variety of different greases lubricating plastic on mild-steel bearings have shown that the choice of lubricant can have a very marked effect on performance.

The traditional recommendation as a grease for plain bearings is a general purpose lime-soap grease (cup grease) of medium–soft consistency. This type of grease has not been found the most satisfactory for marginally lubricated plastic bearings where the lubricant was applied initially only or at widely spaced intervals, and it would not be used where optimum performance was desired.

High-quality lithium-based greases have proved the most successful and reliable type of lubricant under the two types of test used in this evaluation. Thickened silicone oil lubricating quality greases have also given long life in the wear tests and would be particularly suitable for bearings running at elevated ambient temperature (above 70°C), where full benefit of the inertness and good

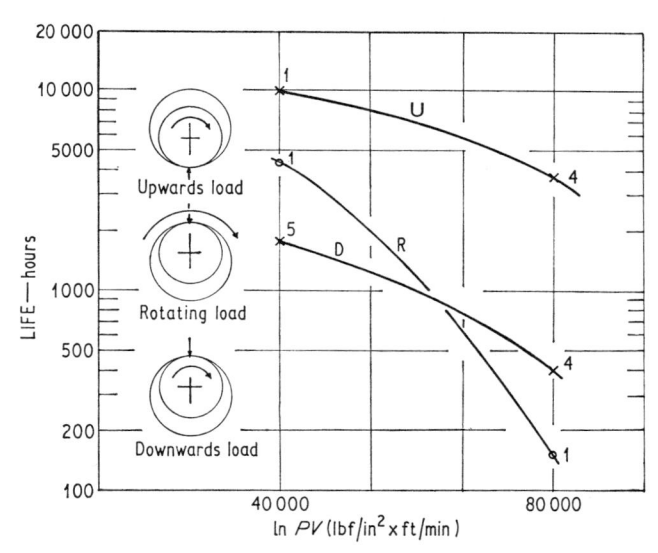

The bushes were run with initial-grease-lubrication-only, at 123 ft/min (0·63 m/s) and with 0·0025-in (0·063-mm) clearance on Beta and Libra test rigs described under 'Performance testing and test machines'. Upward (U), rotating (R), and downward (D) loadings were compared.

The numbers beside points on the graphs give the number of results of which the point shown is the geometric mean.

Fig. 6.4. Effect on performance of type and direction of steady loading on ⅝-in (16-mm) bore Glacier DX bushes

viscosity–temperature characteristics of a silicone oil would be enjoyed.

Small additions of MoS$_2$ to the lubricating greases have proved unharmful; but with more than 10 per cent filler there is evidence of a reduction in performance, except possibly at very low sliding speeds. Graphite-loaded greases are not recommended. As previously mentioned, e.p. additives are not necessary and could conceivably be detrimental to thermoplastic polymer bearing performance. However, there is no conclusive evidence from the wear tests of any attack on the plastic by the chemically active ingredient.

The performance of different greases as lubricants for Glacier DX bearings against mild steel is summarized in Table 6.2. All were readily available brands from various manufacturers. These performance ratings are based on the life of a DX bush running at 40 000 PV (lbf/in^2 × ft/min) (1·4 MN/m^2 × m/s), 125 ft/min (0·62 m/s), against a fine ground En 1A journal with initial-lubrication-only on the Libra machines. Zircon test conditions were 16-lb load on a ½-in wide flat specimen, 1300 rev/min (1·7 m/s), for 16 h.

Greases in the 'not recommended' C category may be used because of some desirable property other than direct beneficial effect on life; and the life may be adequate, provided the application is not severe.

OIL VERSUS GREASE

In addition to making comparisons between the performance of different types of grease, the relative wear and seizure resistances of conventional metal bearing materials, namely white metals, copper, and aluminium based alloys, have been assessed with marginal oil lubrication and against hard and soft mild-steel journals.

The results presented in Table 6.3 give wear-rate ratings based on the actual measured wear rates in inches per hour × 10^5 obtained from Beta rig tests described under 'Performance testing and test machines'. Seizure resistance ratings are based on the number of hours run before seizure occurred, with 100 indicating no seizure in any test where at least two duplicate tests were run.

Detailed consideration of the ratings obtained reveals some divergences from expectation based on field experience, e.g. the excellent performance of oil-lubricated aluminium–zinc has not been supported in engine tests. Nevertheless, the VW 'Beetle' engine has run on solid aluminium–zinc crankshaft main bearing bushes with little trouble for many years.

The oil lubricated tests were run using a 9-min off, 1 min on, stop–start cycle, with loads increased in steps up to 2000 lbf/in^2 (13·8 MN/m^2) and a surface speed of 250 ft/min (1·27 m/s) for up to 336 h. Oil supply rate was 1 drop (30 μlitre) per minute. Diametral clearance was 0·005 in (0·127 mm).

The constant PV continuous running, grease lubricated test conditions were 154 lbf/in^2 (1·06 MN/m^2) load at 130 ft/min (0·66 m/s). Tests were run for up to 336 h with initial-lubrication-only. Diametral clearance was 0·0025 in (63 μm).

The reasoning behind the choice of these test conditions will now be discussed. With an oil-lubricated bearing, wear occurs only when the hydrodynamic film breaks down and direct metal-to-metal contact is made. This would occur on start-up under load during the time before film lubrication was re-established. The 9-min stationary

Table 6.2. The performance of different greases

Ref.	Grease type*	Performance rating† Libra	Performance rating† Zircon	Other details
1	Lime or calcium soap based	D	D	
2	Lime or calcium soap based		C	With e.p. additive.
3	Lime or calcium soap based	C	D	With 12% graphite.
4	Soda or sodium soap based		B	
5	Soda or sodium soap based		B	Synthetic soap for high temperatures.
6	Soda or sodium soap based	A	A	Complex soap with high viscosity, high VI oil.
7	Lithium soap based	A	A	Hydroxystearate base.
8	Lithium soap based		B	With e.p. additive.
9	Lithium soap based	A	A	With 3% MoS$_2$.
10	Lithium soap based		B	With no additives.
11	Lithium soap based	A	B	Silicone oil.
12	Lithium soap based		B	Silicone oil + 10% MoS$_2$.
13	Clay (bentone) based		B or C	
14	Clay (bentone) based		B or D	With 10% MoS$_2$.
15	Clay (bentone) based	B	B	Silicone oil.
16	Lead soap based	D	C	With 20% MoS$_2$.
17	Silica based		D	With 10% MoS$_2$.
18	Rubber grease (lithium hydroxystearate based)	B	D	Castor oil.
19	Carbowax (high MW polyethylene glycol thickened)	B		Polyethylene glycol fluid.
20	P.t.f.e. powder thickened	C		Fluorinated ether.
21	P.t.f.e. powder thickened	C	C	Chlorotrifluoroethylene fluid.

* Where not stated the oil present is a mineral hydrocarbon.
† Details of the test machines are given under 'Performance testing and test machines'. In this table the following ratings are used:
  A = recommended; B = probably acceptable but not yet fully evaluated; C = not recommended; D = unacceptable.

Table 6.3. Comparison between bearing metal performance with oil and with grease lubrication on Beta test

Ref.	Bearing material	Typical indentation hardness, HV	Oil lubricated Wear-rate rating*	Oil lubricated Seizure resistance rating†	Grease lubricated Wear-rate rating*	Grease lubricated Seizure resistance rating†
1	Tin-based white metal (SAE 12)	30	2	100	20	40
2	Lead-based white metal (SAE 15)	20	3	90	20	40 (60)‡
3	Copper 30% lead sinter (SAE 48 approx.)	50	25	40	20	30 (60)
4	Bronze 24% lead sinter bimetal (SAE 799 approx.)	60	40	30	10	50
5	Bronze 10% lead sinter bimetal (SAE 797)	100	60	20	5	100
6	Bronze 26% lead cast (SAE 794 approx.)	60	40	35	2	100
7	Bronze 10% lead cast (SAE 792)	70	60	20	2	100 (100)
8	Phosphor-bronze cast (SAE 65 approx.)	90	100	20		
9	Aluminium 20% tin bimetal	40	30	50	10	70 (70)
10	Aluminium 6% tin bimetal (SAE 770)	50	25	30	5	60
11	Aluminium 5% zinc cast	60	20	60		
12	Aluminium 10% lead bimetal	45	20	55		
13	Aluminium 12% silicon bimetal	60	50	20	20	30 (50)

The explanation of this table is given under 'Oil versus grease'.
* The *lower* the wear-rate rating, the better the wear resistance.
† The *higher* the seizure resistance rating, the better the seizure resistance.
‡ Results in parentheses were obtained against hardened journals.

period and high load were chosen to ensure that the oil would be squeezed out from between the mating surfaces.

Grease-lubricated plain bearings are often run with infrequent re-lubrication, either by design or oversight. Hence, the ability to survive with minimum maintenance is an important factor in the choice of such a material. The load and speed were chosen empirically at a level which would discriminate between good and indifferent bearing materials under conditions which were still realistic.

Most of the ratings in Table 6.3 are based on the average of test results accumulated over several years of running five separate test heads.

The oil was a branded high-quality SAE 5 hydrocarbon hydraulic oil with anti-oxidant, anti-rust, anti-foam, and anti-wear additives. The grease was an NLGI classification 2 to 3 lithium hydroxystearate thickened SAE 10, V.I. 95, hydrocarbon oil with anti-oxidant additive.

The significant feature of Table 6.3 is the partial reversal of the order of performance rating shown, comparing the results for oil lubrication with those where grease was the lubricant. With oil, the soft white metals are indisputably superior to all other metal bearing materials (only in plastics are there materials with comparable or better wear resistance with oil lubrication). Yet if grease is substituted for oil, the white metals become almost seizure prone and subject to relatively high wear rates. Almost the complete reverse appears to be the case with the hard lead bronzes. A different viewpoint is that there is a definite trend, largely independent of material composition, for bush wear resistance and seizure resistance to decrease with increase in hardness and melting point when oil was the lubricant, and to increase with hardness and melting point with grease lubrication.

Although the loads, speeds, and procedures for the oil lubricated tests were quite different from those for the grease tests, previous work (not reported) had shown similar trends from identical tests with initial oil lubrication instead of grease. This indicates that the difference in performance stemmed solely from the nature of the lubricant employed.

An explanation for this apparent anomaly, for oil is usually considered to be the active lubricating phase in both cases, is based on the much greater resistance to shearing, particularly in very thin films, of the grease resulting from the fibrous nature of the thickener. This resistance to shearing would lead to the generation of heat at the bearing interface, possibly sufficient to cause local melting of the low melting point phase of the bearing. In the case of white metals this low melting point phase is also the load supporting matrix, so any loss due to melting and wiping would lead to measurable wear.

Harder, higher melting point bearing materials would be less affected by the higher lubricant temperature and would benefit from the superior boundary lubricating ability of the adherent grease film.

The few test results against hardened journals (shown in Table 6.3 in parentheses) suggest the desirability of using a hard shaft even with the soft white metals when grease was the lubricant.

THE IDEAL GREASE FOR PLAIN BEARINGS

The term 'grease' is used to cover a wide range of products, from near liquids to hard solids, which all consist essentially of a lubricating fluid thickened with a solid, with possibly small additions of additives (molybdenum disulphide, graphite, e.p., rust preventive, anti-oxidant).

Sufficient is now known about the techniques of the manufacture of greases to cater for special circumstances, e.g. high-temperature greases which are formulated with temperature-resistant fluids (polyphenyl ethers, diesters, silicones) and solids (oleophilic clay, silica, dye, graphite, acetylene black). Special greases can be quite expensive. The ratio of cost of hydrocarbon oil: diester: silicone-based greases is about 1:10:100. For reasons of cost and availability, general-purpose greases are soap-gelled hydrocarbons, the consistency of which can be controlled by choice of oil (high or low viscosity, naphthenic or paraffinic), proportion and type of soap (e.g. sodium, calcium, lithium), and the blending technique.

The ideal grease for plain bearings should be insensitive to shear, temperature, and oxidation.

The high shear condition between the journal and the bearing is likely to cause a breakdown of the grease. The loss of consistency of soap-based grease in bushes is due either to a separation of the two phases or to shear breakdown of the filamentary structure of the soap. Soap fibres in the gel structure of the grease have been observed by the electron microscope to be about 100 μm long and 5 μm in diameter. It is interesting to speculate on the properties of a grease gelled with inorganic or carbon graphite fibres, of a similar micro-size range, if these were available. Would such fibres, although more temperature resistant, be less sensitive to shear? (The shear breakdown of soap may be largely a thermal process.) Would this type of graphite act as a lubricant and prolong the life of the bearing? Asbestos has been used as a macro-size fibre additive to calcium-based grease for use in wartime amphibious tanks; but the aim was to increase the ability of the grease to resist water erosion rather than to act as a primary gellant. Shear breakdown of soap fibres may be analogous to the shear degradation of viscosity index improvers in oil or the reduction in molecular weight of strongly sheared polymers. Susceptibility depends upon the rate of shear and the length-to-diameter (l/d) ratio of the fibre. The process explains the rapid breakdown of polyethylene- and polypropylene-thickened greases (these plastics have molecules with a l/d ratio greater than 1000:1) and the insensitivity to shear of the high-temperature gellants, which are globular or plate-like particles (l/d ratio about 1:1).

Unreported experiments have shown that polyethylene glycol fluids act as useful lubricants for plastic bushes. However, a grease-like aqueous solution of high molecular weight solid polyethylene glycol was a poor lubricant because the shear forces operating between the journal and the bearing rapidly degraded the PEG molecules. The grease was transformed to a watery consistency with no lubricating properties. The interest in this unusual lubricant arose from publicity given to its use in the reduction of skin friction of ships' hulls.

Temperature considerations are complex. At low temperatures the liquid component must not freeze; for high-temperature operation the liquid component must not evaporate, degrade chemically, or oxidize, and the gelling component must not melt. Gellants resistant to high temperatures have been mentioned earlier. Polytetrafluorethylene (p.t.f.e.) is quite unlike previous high-temperature gellants. Greases containing the low-friction plastic solid p.t.f.e., both as an additive and as a gellant, have recently been introduced. Our experience with p.t.f.e. as a sole gellant is that the grease is unstable, and an oil layer soon separates on the surface of the grease during storage. Using a high-viscosity hydrocarbon base oil and a micro-crystalline hydrocarbon gellant, storage-stable p.t.f.e.-filled greases have been made which have useful low-friction properties in large bearings. However, our testing has shown that the anti-wear properties of this p.t.f.e. grease are unexceptional.

For the normal temperature operation of a bearing which has been initially greased only, the liquid component must remain stable over long periods. Naphthenic may well be preferred to paraffinic oils because the response to anti-oxidants is greater. Greases based on a high viscosity will be preferred to those based on low-viscosity oils, except for low-temperature environments, because the rate of evaporation is lower and the load-supporting ability is better (thoroughly sheared grease is, effectively, base oil). Sims and Hunt (6) demonstrated the excellent high-load, low-speed characteristics of greases containing high-viscosity oils (SAE 40 and above) compared with those of one made with low-viscosity (SAE 20) oil. They also demonstrated that a grease based on a low-viscosity oil could be upgraded by the addition of 25 per cent bitumen. It is not clear whether the improvement resulted from the increased viscous consistency of the grease or whether the bitumen was acting as a crude extreme-pressure additive; or, indeed, whether differences between the soap or oil components of the greases were responsible. It is known that low volatility is not the sole required property because silicone greases (silicone oil with thickener) have given no better performance than the standard lithium-based hydrocarbon grease in plastic bearings. In addition, greases based on fluorocarbon fluids have performed badly as plastic bush lubricants.

It is not known why the best greases for plain bearings are successful, but it is clear that wear results are influenced at least as much by surface chemistry of the bearing and journal as by the rheological behaviour of the grease.

CONCLUSION

The development and wider use of grease as a lubricant for plain bearings have been hindered by the lack of a theoretical basis from which designers could work.

Where greased plain bearings are used with infrequent relubrication, a high-quality NLGI classification 2 or 3 lithium hydroxystearate based grease with a plastic-lined bearing material has proved the most successful.

With marginal oil lubrication, soft tin and lead based bearing materials gave the best wear and seizure resistance. Under marginal grease lubrication conditions, the ranking was practically reversed; hard lead bronzes proving the best materials, while white metals tended to wipe.

It is probable that there remains considerable scope for the development of greases to suit plain bearing operating conditions.

APPENDIX 6.1

REFERENCES

(1) COHN, G. and OHREN, J. W. 'Film pressure distribution in grease-lubricated journal bearings', *Trans. Am. Soc. mech. Engrs* 1949 (July), 555.

(2) MILNE, A. A. 'A theory of grease lubrication of a slider bearing', *Proc. Second Int. Congress on Rheology* 1954, 427 (Butterworths, London).

(3) CHAKRABARTI, R. K. and HARKER, R. J. 'Frictional resistance of a radially loaded journal bearing with grease lubrication', *Lubric. Engng* 1960 (June), 274.

(4) OSTERLE, F. *et al.* 'The rheodynamic squeeze-film', *Lubric. Engng.* 1956 **12** (No. 1), 33.

(5) SLIBAR, A. and PASLAY, P. R. 'On the theory of grease-lubricated thrust bearings', *Trans. Am. Soc. mech. Engrs* 1957 (August) **79** (m-6), 1229.

(6) SIMS, R. B. and HUNT, R. T. V. 'Design of grease-lubricated plain journal bearings for a crank-operated cold plate shear', *Proc. Instn mech. Engrs* 1964–65 **179** (Pt 3D), 12.

(7) BRADFORD, L. J. *et al.* 'Grease lubrication studies with plain journal bearings', *J. bas. Engng, Trans. Am. Soc. mech. Engrs* 1960, Paper 60-Lub-5.

(8) PRATT, G. C. and WILSON, W. H. 'The performance of steel-backed acetal copolymer bearings', *Wear* 1968 **12**, 73.

Paper 7

A SURVEY OF ROLLING-BEARING FAILURES

A. W. Morgan*   D. Wyllie†

The condition of approximately 600 suspect rolling bearings from electrical machinery was examined. This number did not represent a high failure rate in relation to the number in use at the time, but afforded a basis for the assessment of the frequency of the various types of failure. The grease in use was the R.N. multi-purpose grease XG-274 which consists of mineral oil and a gelling agent (currently a metallic soap) and oxidation and corrosion inhibitors, but some bearings still contained earlier greases. The most frequent causes of failure were corrosion, various machine and fitting defects, and dirt. The defects detected are discussed.

INTRODUCTION

THE INTERPRETATION of damage in rolling bearings is well documented (1)–(6)‡ but there is a shortage of surveys of the incidence of the various defects. A few years ago the conventional uninhibited soda base grease, which had for many years been in use in grease-lubricated electrical machinery in the Royal Navy, was replaced by the multi-purpose grease XG-274 (7). In discussions as to what, if any, follow-up work was required it became apparent that there was considerable uncertainty regarding both the number and the cause of bearing failures. The Ship Maintenance Authority (S.M.A.) had analysed ships' reports from April 1964 to April 1966, but these were not accompanied by details of the actual state of the bearings after removal, and many failures were not reported.

About the same time, all suspect rolling bearings in the electrical machinery in two frigates had been sent to A.O.L. and examined for a period of two-and-a-half years after the ships were refitted. From the small numbers involved no reliable conclusions could be drawn, but it appeared that corrosion, machinery, and fitting faults accounted for a considerable number of the failures and that the overall failure level might be around 3 per cent per annum. It was decided to conduct a trial in which as many ships as possible would take part. All suspect bearings were to be sent to A.O.L. with reports of the circumstances, on forms supplied for the purpose, to A.O.L. and to S.M.A. The trial commenced in late July 1966 with the intention of collecting at least 500 bearings, and was ended a year later with a total collection of 614 bearings, 596 of which were confirmed as having failed in some degree.

METHODS OF EXAMINATION

In order to ascertain the cause(s) of failure, it was nearly always necessary to completely dismantle the bearings so that all the internal surfaces could be carefully examined. The only exceptions were bearings which were already dismantled on receipt because one or more parts had fractured, and ball bearings that were so badly corroded that the balls could not be moved after removal of the cage.

Before dismantling, each bearing was tested for freedom of movement (*a*) in its as-received condition, and (*b*) after complete removal of the grease, a thorough cleaning with solvents, and lubrication with light mineral oil. The quantity of grease was noted and a representative sample was taken and examined for identity, consistency, colour, and presence of foreign matter. The dismantling procedure usually necessitated the drilling away of the heads of the rivets holding the two cage halves together, and great care was taken to wash away the drillings before removing the rolling elements so that these and the tracks were not scored or indented. Before and after examination the bearing parts were kept in a sealed container in which a piece of vapour-phase inhibited paper was placed to prevent the parts corroding.

Most defects could be observed without the aid of a microscope, but the instrument was necessary occasionally to detect small scores and indentations and light corrosion, especially in small bearings. A binocular instrument was used which had a good depth of field and a wide range of magnification, though it was rarely necessary to use magnifications higher than 15 or 25. The detection of a

---

The MS. of this paper was received at the Institution on 24th July 1969 and accepted for publication on 14th October 1969. 33
* *Experimental Officer, Admiralty Oil Laboratory, Fairmile, Cobham, Surrey.*
† *Principal Scientific Officer, address as above.*
‡ *References are given in Appendix 7.1.*

spiral ball path or one with lobes (indicating out-of-roundness) was facilitated by spinning the race on a turntable.

The actual assessment was in accordance with the authorities listed (2)–(7). The nature of the ball paths indicates the presence or otherwise of defects such as misalignment, out-of-roundness, rotating radial load, and excessive axial load; fretting corrosion and/or polishing of the outer diameter and bore often match up with peculiarities in the ball path. False and true brinelling and passage of electric current produce characteristic marks on the tracks, and the presence of abrasive dirt results in scores on the rolling elements and indentations in the tracks.

Sometimes scores were present in bearings although the dirt which caused them was no longer present. In a few cases the bearing contained ample fresh grease but showed symptoms of having run with insufficient grease. It is probable that an attempt may have been made to cure a noisy bearing by putting in new grease and the bearing had been replaced when the remedy proved ineffective.

The majority of bearings were sent in because they appeared to be causing noise and/or vibration, and all the bearings considered to have failed had some defect(s) which could have produced these effects. Several of the failed bearings could have continued to run, but only with an unacceptable degree of noise and suffering from defects which would promote vibration. This in turn would have accelerated surface failure and complete breakdown of the bearing.

THE GREASES

Specification DGS/6921 governing supplies of XG-274 was first issued in 1963. This contained a typographical error and a corrected version was issued as DGS/6921A a few months later. The first approval of the specification was given in 1964 to a grease which had been in use on a trial basis since 1962. Up to the end of the trial only one XG-274 grease, a lithium base grease, had been purchased and issued. It was found in 439 of the 614 bearings received. It takes time for any new material to come into general use and for all older types to be replaced.

A soda/lime base grease, generally of the required performance standard, was purchased and issued on a trial basis from 1960 to 1963. It tended to harden in contact with water and did not meet the rust prevention test in DGS/6921. Of the bearings collected, 105 arrived after running on this grease, referred to herein as prototype XG-274, while 15 bearings contained both the prototype XG-274 and grease purchased against the specification; 8 bearings still contained old type grease in use before the prototype XG-274, and the grease in 14 bearings could not be identified.

No less than 34 bearings contained no grease on receipt. Of these, 16 had been drawn from store, cleaned out for use, and sent on as suspect; 3 arrived still in preservative; and the remaining 15 were taken from machines in use, and in some instances had been washed out and looked at before dispatch to A.O.L.

THE FAILURE PATTERN

Fig. 7.1 shows the failure pattern in terms of time since the bearings were fitted. It will be seen that the highest failure rate was in the first three months: most of this did in fact occur in the first month. Thereafter the number of failures decreased overall, except for a marked peak after three years which coincides with the time ships tend to come in for refits and routine checking of equipment.

The failure pattern in terms of actual running hours is given in Table 7.1, distinguishing on this occasion between XG-274 to specification DGS/6921A and the other greases. Again the failure rate is highest initially. It was found that a few bearings were unserviceable when unpacked for use after storage, and a number failed on test. The other greases had few early failures but these would have occurred before the start of this trial when XG-274 was not in general use.

Ships were unable to state the time since fitting or the running hours for about one-fifth of the bearings. As

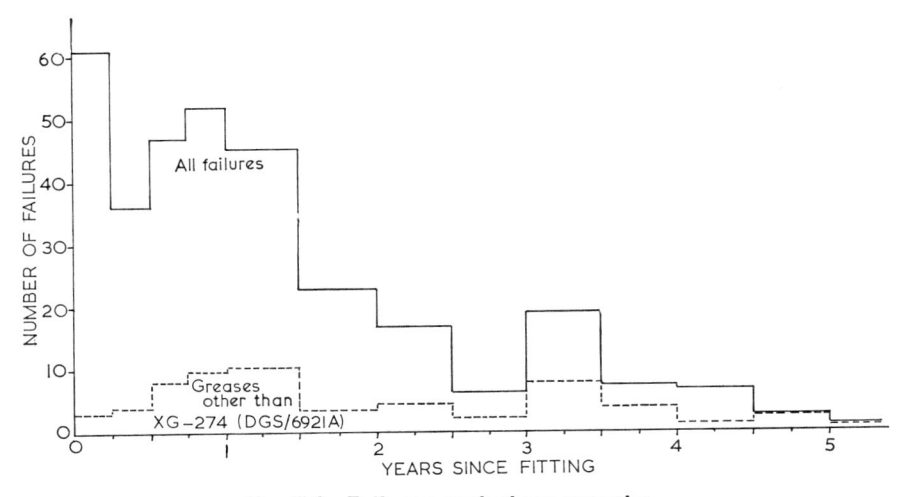

Fig. 7.1. Failures each three months

would be expected this occurred most frequently with bearings containing the older greases.

Table 7.1. Running hours to failure

Running hours to failure	Bearings lubricated with XG-274 to DGS/6921A	Other bearings	Totals
Unused	0	14	14
On test	32	0	32
After test up to 500	30	14	44
501– 1 000	18	4	22
1 001– 2 000	44	13	57
2 001– 3 000	19	4	23
3 001– 4 000	18	3	21
4 001– 5 000	33	7	40
5 001– 6 000	24	13	37
6 001– 7 000	11	1	12
7 001– 8 000	18	7	25
8 001– 9 000	16	3	19
9 001–10 000	26	7	33
10 001–12 500	12	2	14
12 501–15 000	11	3	14
15 001–17 500	7	4	11
17 501–20 000	10	2	12
20 001–25 000	12	6	18
25 001–30 000	8	6	14
Over 30 000	2	5	7
Not known	75	52	127
Totals	426	170	596

Of the bearings received, 525 were ball bearings and 89 were roller bearings. The great majority of both were rigid single-row with various types of cage. None was lubricated-for-life.

The total bearings adjudged to have failed in some degree should be set against a probable 39 000 bearings at risk in the ships taking part in the trial. This gives a failure rate of 1·5 per cent per annum. From comparison with other records S.M.A. was certain that by no means all the failures had been sent in. The probable return was considered to be at most 60 per cent and at least 40 per cent of all suspect bearings. In the light of this the failure rate is likely to have been at least 2·5 per cent but not more than 4 per cent per annum.

The main causes of failure were corrosion followed by dirt, misalignment, insufficient grease, and excessive axial load (Table 7.2). No other defect was listed as a main cause of failure in as much as 10 per cent of the bearings, several in only 3 per cent or less. The number that failed owing to some form of machine or fitting defect was almost the same as the number that failed through corrosion.

A few defects appeared to be rather more prominent in bearings lubricated with XG-274 supplied and approved against specification DGS/6921 than in all bearings regardless of how they were lubricated. These were dirt, misalignment, and true brinelling (damage by blows and unduly forceful fitting or removal). The first may be fortuitous, the others are fitting defects likely to give comparatively early failures and therefore to be less prevalent

Table 7.2. Summary of main causes of failure

Condition	All bearings Number	All bearings Percentage of bearings	Bearings lubricated with XG-274 to DGS/6921A Number	Bearings lubricated with XG-274 to DGS/6921A Percentage of bearings
Bearings received*	614	—	439	—
Bearings failed	596	97	429	98
Bearings badly failed	307	50	209	48
Bearings too badly damaged for cause of failure to be assessed	18	3	10	2
Bearings with fractured components	39	6	28	6
Bearings flaked	180	29	129	29
Bearings with severe corrosion	77	13	48	11
Bearings with the following main causes of failure (more than one can be present in one bearing)				
Corrosion	238	39	151	34
Dirt	112	18	91	20
Misalignment	74	12	59	13
Insufficient grease	72	12	56	13
Excessive axial load	63	10	43	10
Hard grease	43	7	5	1
False brinelling	38	6	27	6
True brinelling	21	3	20	5
Grease inadequate	19	3	14	3
Natural fatigue	19	3	17	4
Poor fit	19	3	12	3
Rotating radial load	10	2	7	2
Soft grease	4	<1	3	<1
Overheating	4	<1	4	<1
Passage of electric current	3	<1	2	<1
Too much grease	1	<1	1	<1

* It is estimated that the bearings received represent 1·5 per cent of the bearings at risk. It is believed that the return might have been 2·5–4 per cent if all suspect bearings had reached A.O.L.

in the bearings which were still lubricated with the older greases.

XG-274 was much less likely to give trouble from hardening than the older greases, which is to be expected as a special clause to control the hardening with water experienced with the prototype XG-274 was included in specification DGS/6921. There was also an improvement in respect of corrosion.

Many bearings suffered in various degrees from flaking of the tracks and, occasionally, of the balls or rollers. A study of the main causes of failure in these bearings showed that the most frequent was excessive axial load followed by corrosion, insufficient grease, misalignment, and natural fatigue in that order. Corrosion and dirt between them totalled 50 as against 54 for excessive axial load and 102 for all machine and fitting defects.

CORROSION

This fault, which was prevalent in the bearings received at A.O.L., could have been influenced by the way the bearings were handled on removal, how they were kept prior to dispatch, and the manner in which they were packed. This was kept in mind and all were examined for scores, indentations, and imbedded particles resulting from running a bearing in which corrosion was sufficient to contribute significantly to failure. Surface corrosion on external parts was noted, but this is not included here and was not reported to S.M.A.

Table 7.3 is set out in terms of grease type and severity of corrosion. Severe corrosion will almost always have occurred before the bearing was suspect, and because of the method of examination the other corrosion failures listed should also be genuine. This is not as certain with the considerable number that had only slight or very slight corrosion, including many with very minor defects.

It will be seen that the percentage of corrosion failures is appreciably lower with XG-274 than with other greases. In general, the rust inhibitors in XG-274 have reduced the severity of attack so that many which would probably have been failures with the older greases were only affected to a slight or very slight extent. However, one-third of the bearings lubricated with XG-274 that were sent to A.O.L. as defective had corrosion as a main cause of failure. XG-274 already meets one of the most severe rust-prevention tests in use and it would not be easy to obtain or specify a higher standard.

Some bearings were reported as having defective seals, but as the majority were described as 'seals not fitted' it is doubtful whether it is justifiable to take much notice of those that were defective. If the corrosion failures with defective seals are excluded, then the failure rate is reduced from 34 to 31 per cent for approved XG-274 and from 49 to 43 per cent for the older greases.

It was concluded by S.M.A. that the type of enclosure appeared to be irrelevant to failure, the failure being in proportion to population, the majority of the machines affected being sited in areas of high humidity or where water and steam have easy access.

The failure rate is highest with those bearings which, on arrival, contained no grease. This is not surprising. However, they include some bearings which had been unpacked from store and rejected before fitting, almost certainly because the fitter saw marks and corrosion. They would otherwise have been installed in machines and might have failed on test.

Of the 72 bearings received with insufficient grease as a main cause of failure, 50 showed some signs of corrosion but only in 12 of these was the corrosion extensive enough to be considered an additional main cause of failure. In other words, in only 12 of the 238 bearings with significant

Table 7.3. Incidence of corrosion

Condition	Lubricated with XG-274 to DGS/6921A Number	Lubricated with XG-274 to DGS/6921A Percentage of total on XG-274	Lubricated with other greases Number	Lubricated with other greases Percentage of total on other greases	Unused bearings, suspect when taken for fitting	Used but no grease present	Totals
Severe failures	48	11	21	15	1	7	77
Other failures	103	23	48	34	10	0	161
Total failures	*151*	*34*	*69*	*49*	*11*	*7*	*238*
Slight or very slight corrosion	205	47	52	37	8	5	270
Free from corrosion	81	18	20	14	0	1	102
Too damaged to assess	2	0·5	0	0	0	2	4
Total received	439	—	141	—	19	15	614
Defective seals reported:							
Bearings failed from corrosion	13	3	9	6	—	1	23
Bearings not failed	4	1	3	2	—	0	7
Bearings water-marked:							
Failed from corrosion	44	10	11	8	5	3	63
Not failed	27	6	1	1	1	1	30

Fig. 7.2. 'Water-marking' at ball positions on bearing from ventilation-fan motor, indicating that corrosion occurred while bearing was stationary

corrosion could lack of grease be a possible reason for the corrosion.

In some bearings the corrosion pattern corresponded to various extents with the position of the balls/rollers when stationary (Fig. 7.2). This 'water-marking' must have occurred while the bearing was stationary. It was present in 5 of the 11 bearings which were corrosion failures when unpacked for use, and in 58 of the 227 corrosion failures of bearings in installed machines. A significant number must have suffered at least in part while idle in moist conditions.

MACHINE AND FITTING DEFECTS

The total number of machine and fitting defects severe enough to be regarded as main causes of failure was 271 from 232 bearings. Thus, about one in three of the bearings failed for reasons over which the grease could have little control. A much larger number had minor defects in this category.

These are summarized in Table 7.4 commencing with dimensional defects, followed by machine defects and fitting defects, and finishing with defects that cannot easily be assigned to any one category. It is realized that some of the failures attributed to out-of-round shafts or housings may have been the result of distorting the races by uneven tightening of the end caps, i.e. they may really have been 'fitting defects'. Also, misalignment of races is not necessarily the fault of the fitter as the same effect can result from deflection of the machine parts and also incorrectly machined housings and abutments.

A considerable number of bearings showed fretting corrosion to some degree, but with many the internal condition of the bearing rather than poor fit was responsible for the bearing moving on its shaft or in its housing. A few bearings were such a poor fit in their housings that

Table 7.4. Machine and fitting defects

Defect	A main cause of failure	Not a main cause	Totals
Dimensional defects			
Shaft under size (too slack)	1	80	81
Shaft over size (too tight)	1	0	1
Shaft out-of-round	6	2	8
(including shaft 3-lobe)	2	1	3
Housing over size (too slack)	19	98	117
Housing under size (too tight)	3	0	3
Housing out-of-round	14	6	20
(including housing 3-lobe)	0	2	2
Machine defects			
Rotating radial load	10	7	17
Passage of electric current	3	2	5
Overheating (not caused by bearing defect or failure)	4	0	4
Fitting defects			
Inner misaligned	38	55	93
Outer misaligned	47	36	83
True brinelling (including scoring in roller bearings)	21	16	37
Various defects			
Excessive axial load	63	9	72
Excessive journal load	1	1	2
False brinelling	38	133	171

Fig. 7.3. High polish and wear on outer diameter caused by rotation of race in oversize bearing

the outer diameters were highly polished by the rotation (Fig. 7.3).

There were few dimensional defects severe enough to be a main cause of failure. As well as undersize shafts and oversize housings they include occasional too-tight fits and out-of-roundness in shaft or housing which showed up in varying ball path width and fretting patterns. Minor examples of slack fits were fairly common.

Genuine machine defects were uncommon. There were occasional instances of rotating radial load caused by out-of-balance forces, passage of electric current, and overheating from sources outside the bearing. Many bearings showed signs of overheating, but as these were failures in some degree from other causes it was concluded that overheating was the result of other defects, not a cause.

Fitting defects occurred as a main cause 106 times—or 95 if we include, as one only, bearings with serious misalignment of both races. The relative occurrence of misalignment of the inner and the outer race is shown. Some bearings were interesting exhibits with a ball path which went from side to side on one race, combined with a ball path of unusual width on the other. Roller bearings with tilted outer races often had roller paths similar to that shown in Fig. 7.4.

In addition to false brinelling, which is widely understood, we have used the term 'true brinelling'. False brinelling is the damage caused by vibration when the bearing is stationary, and is so called because the damage resembles the marks produced in the Brinell test for hardness. True brinelling is damage resulting from blows, often during forcible fitting or removal of the bearing. They show up as ball or roller impressions on the tracks and as damage to the rolling elements themselves. A typical example is the fitting of a ball bearing on a shaft by applying force to the outer race rather than to the inner, resulting in deformations on the edges of the tracks at each ball position. Under this heading we have included scoring and scuffing of the tracks and rolling elements of roller bearings caused by misapplied force during fitting or removal.

True brinelling was evident in 18 ball and 19 roller bearings and this was a main cause of failure in 10 ball and 11 roller bearings. But 529 ball bearings were received and only 89 roller bearings. Although it may be a minor factor (2 per cent) in failures of ball bearings, damage on assembly (or perhaps on removal) was of importance in 12 per cent of the roller bearings.

False brinelling was a significant factor in the failure of 38 bearings and was light or very light in considerably more. A very interesting example occurred in the earlier trial on a bearing from a turbine turning motor which must have vibrated when the turbine was running. Other examples came from machines such as compressors where several were fitted and those which were in use vibrated the ones standing by.

The third most common cause of failure was excessive axial load. This has been taken to be the failure in bearings under thrust load when the ball path is not only displaced in consequence, but has suffered heavy wear (Fig. 7.5). This type of wear was detected in 72 bearings and since mild examples would not normally be detected it is not surprising that in 63 it was considered a main cause of failure. One-third of the bearings that failed from excessive thrust load contained balls which had been rotating in one plane under more than the load they were intended to carry and showed wear bands.

An initial assumption was that excessive axial load would be most prevalent in vertically mounted machines. In fact the failures mounted vertically turned out to be almost the same in number as the failures mounted horizontally. Of the total bearings received, however, the ratio of horizontally to vertically mounted was

Fig. 7.4. Roller path on tilted outer race, consisting of two lines, overlapping at two places, made by the edges of the rollers

Fig. 7.5. Ball path completely to one side of inner race track, showing advanced flaking as result of excessive axial load

approximately 7:3. Therefore, it would appear that the vertically mounted machines are more prone to this defect.

Other main causes of failure which affected these 72 bearings were insufficient grease in 10, misalignment in 9, hard grease in 8, corrosion in 6, out-of-round housing in 3, and dirt in 1. Without these contributory factors there would undoubtedly have been less failures attributed to excessive axial load.

DIRT AND GREASE FACTORS

Dirt was present to some extent in 70 per cent of the bearings (see Table 7.5). It may be that in some bearings this dirt got in after the bearing was taken out and before it was packed for dispatch to A.O.L. Dirt as a main cause of failure affected almost half the number of bearings which were affected by corrosion and was the second most common cause of failure. It was not regarded as a main cause of failure unless scores and/or indentations were present on the balls, rollers, and/or tracks (Fig. 7.6).

Although XG-274 as supplied is very clean, the percentage of bearings lubricated with XG-274 which contained dirt was higher than the percentage of bearings containing dirt and lubricated with other greases. This is a reflection on the difficulties of keeping dirt out during fitting and of preventing it getting into grease tins which are opened and shut in working spaces.

The grease in 102 bearings was unduly hard. All but

Table 7.5. Dirt and grease factors

Factor	XG-274 DGS/6921A	Prototype XG-274	Others	Totals
Dirt present	327	80	21	428
Dirt a main cause of failure	91	15	6	112
Grease hard	8	93	3	104
Hard grease a main cause of failure	5	38	0	43
Grease insufficient	57	5	13	75
Insufficient grease a main cause of failure	56	4	12	72
Grease inadequate	17	6	0	24
Inadequate grease a main cause of failure	14	5	0	19
Grease too soft	3	1	0	4
Soft grease a main cause of failure	3	1	0	4
Too much grease	1	0	0	1
Too much grease a main cause of failure	1	0	0	1
Number of bearings with each grease	439	104	71*	614

* Includes 19 unused bearings.

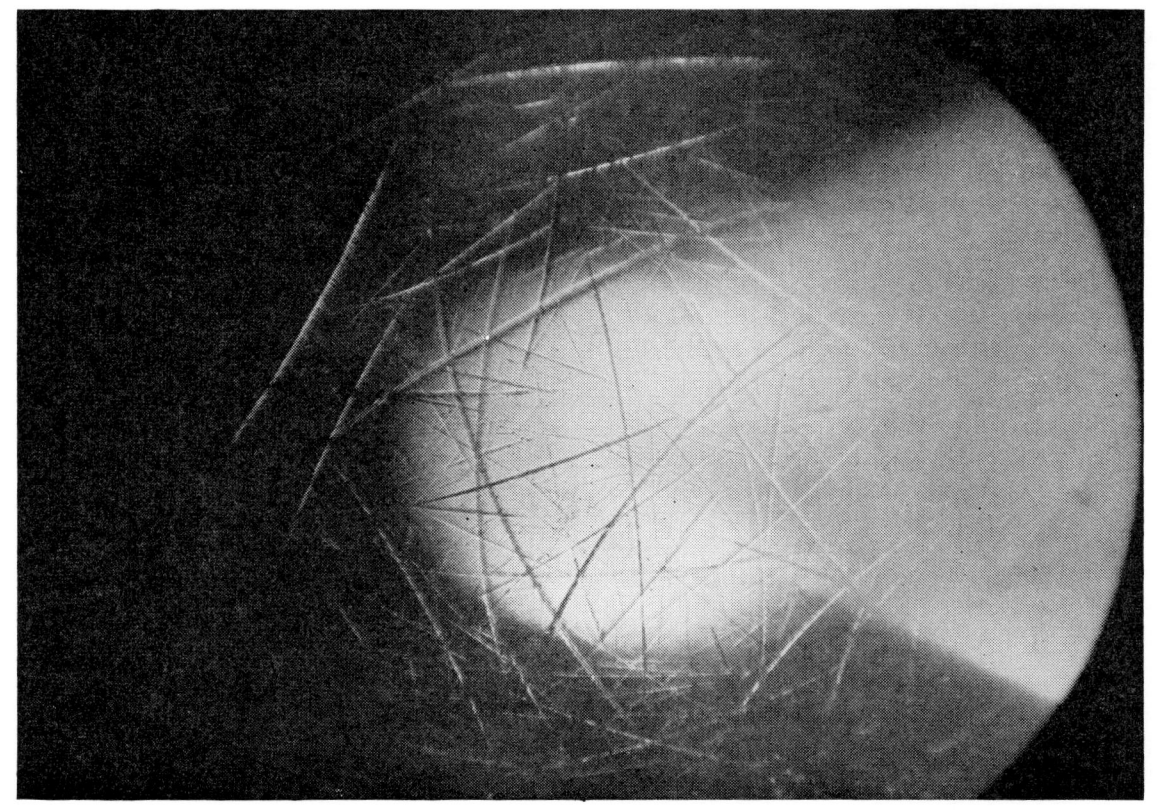

Fig. 7.6. Deep scores on ball (magnification ×15) caused by dirt

11 of these contained the prototype XG-274 which was liable to harden in presence of moisture. In the majority, hard grease was not a main cause of failure of the bearing. A grease that was too soft was seldom noted.

Insufficient grease in the bearing was the fourth most common cause of failure. This may be a result of the practice of not relubricating in the hope of reducing failures from dirt or from wrong greases introduced during relubrication. Only one bearing appeared to have failed because of too much grease.

A small number of bearings appeared to have failed owing to inadequate lubrication, there being considerable wear with no other apparent explanation. Nearly all failed because of cage wear. More than half were large ball bearings with machined brass cages centred on the inner race, with heavy or very heavy wear on the inner diameter of the cage, sometimes to the extent that the cage had dropped. Most of the remainder had pressed steel cages showing significant wear in the pockets, and bearings in which the shape of the pockets resembled a bow-tie seemed more prone to wear.

A count of the number of bearings of these two types on the ships in the trial revealed that the failure rate was not significantly greater than with other bearings, but it did tend to show up as cage wear.

It was noted earlier that 10 of the 63 bearings listed as having excessive axial loading also contained insufficient grease. It is possible that there would have been less failures attributable to excessive axial loading if regular relubrication had taken place. The total number of bearings affected to the extent of failure by excessive axial load or insufficient or inadequate lubrication was 144, i.e. 23 per cent of the bearings received. Some of these failures might have been prevented by regular relubrication.

DISCUSSION

It has been shown that the failure rate is small and, in consequence, the frequency of the minor causes of failure is insignificant. It is gratifying that dimensional defects leading to genuine cases of loose or poor fit, and machine defects such as serious out-of-balance or passage of electric current, were rare.

Defects which were important in 10–13 per cent of the bearings received, equivalent to only between 0.25 and 0.5 per cent of the bearings at risk, were excessive axial load, misalignment, and insufficient grease. The first two probably reflect the difficulties of fitting machines in confined spaces. The last may be a result of the current practice of running bearings in R.N. electrical machinery without relubricating. This was introduced to reduce the risks of over-lubricating with the wrong grease or admitting dirt in the process. It would appear that there are some machines which would benefit from regular relubrication.

Despite this it was disappointing that 70 per cent of the bearings received contained dirt to some extent, although in only 20 per cent was this sufficiently important to be a main cause of failure. Bearings have to be fitted in confined spaces under working conditions and the grease is supplied in tins which will be opened and shut a number of times. There is clearly much to be said for the use of grease containers of minimum size and/or tubes in place of tins.

It has been shown that even with a top-quality inhibited grease, corrosion as a main cause of failure occurred in one-third of the bearings, which corresponds to around 1 per cent of the bearings at risk and lubricated with this grease. While one would like to reduce this it would appear that at the moment there is no real prospect of doing so to any significant extent by increasing the rust-preventing properties of the grease.

Water-marking in a number of bearings and wear tracks superimposed indicated that it was not uncommon for corrosion to occur while the machine was idle or, on occasion, before the bearing had even been fitted. It was suspected that some of the corrosion had occurred before machines had started to run, perhaps when bearings were cleaned and not immediately fitted and lubricated.

None of the bearings was lubricated-for-life, which is not surprising as few such bearings were fitted at the time. It may be that an increased use of sealed and lubricated-for-life bearings would help to reduce the failure rate by reducing the opportunities for moisture and dirt to gain access.

ACKNOWLEDGEMENTS

The authors are indebted for permission to publish this paper to the Ministry of Defence (Navy) who are not, however, responsible for the opinions expressed. They also acknowledge assistance given by the Ship Maintenance Authority, the co-operation of ships' staff in supplying the bearings and their histories, and advice received from the members of the grease panel of the Navy Department Fuel and Lubricants Advisory Committee, especially Mr A. W. Humphreys of Skefko and Mr E. A. Goodchild of Hoffmann, and his colleagues Mr F. R. Absolom and Mr F. W. Lambert. Much of the actual examination of the bearings was carried out by Mr D. P. Pailthorpe of A.O.L.

APPENDIX 7.1

REFERENCES

(1) *Interpreting service damage in rolling type bearings* 1953 (American Society of Lubrication Engineers).
(2) *S.K.F. bearings—mounting and maintenance* (Skefko Ball Bearing Company Ltd).
(3) RIDDLE, J. *Ball bearing maintenance* 1955 (University of Oklahoma Press).
(4) WILCOCK, D. F. and BOOSER, E. R. *Bearing design and application* 1957 (McGraw-Hill).
(5) GOOD, W. R. and GUNST, A. J. 'Bearing failures and their causes', *Iron Steel Engr* 1966 (August), 83.
(6) MORTON, H. T. *Anti-friction bearings* 1954 (Morton Bearing Co.).
(7) WYLLIE, D. and JONES, J. T. 'XG-274. A multi-purpose grease for R.N. use', *J. Inst. Petrol.* 1965 **51**, 53.

Paper 8

RHEOLOGICAL BEHAVIOUR OF A NEW HIGH-TEMPERATURE SYNTHETIC GREASE

A. E. Yousif* K. D. Bogie†

The thermal stability of greases has become an important factor in modern mechanical engineering applications. With this in mind, experiments were conducted with a carbon black as a filler in a carrier fluid of siloxane (MS 550) in order to obtain higher operating temperatures. It was determined that the colloidal system formed depended upon solids concentration by weight. Viscosity–shear rate characteristics were observed to be highly non-Newtonian for high solid concentrations and less marked for lower concentration, resulting in poor mechanical stability. Special surface treatment of the carbon black showed an improvement. When subjected to oscillatory shearing of variable strain amplitude the system was again observed to depend upon the percentage solid concentration. The fundamental wave suggests stronger elastic properties at low strain amplitude than for similar systems tested for frequency dependence.

INTRODUCTION

THE RHEOLOGICAL BEHAVIOUR of greases in a lubricated bearing is a complex problem. This is because the grease within the actual contact zone, whose function is the lubrication of the bearing working surfaces, experiences high contact pressures, while the grease in and around the bearing which is outside the contact zone is at ambient pressure (usually atmospheric). The grease outside the contact zone should act as a reservoir for the contact zone, conducting heat away from the zone, and inhibiting the ingress of foreign particles to the zone. The grease outside the contact zone is subjected to shearing and vibration during the normal operation of most engineering bearings.

It was hoped that when the grease was subjected to oscillatory shearing under atmospheric pressure, some knowledge of its behaviour would be obtained which would assist in the solving of the problem.

Greases are known to be non-Newtonian two-phase colloidal systems having the suspension in the carrier fluid. When subjected to oscillatory shearing of a certain amplitude, they exhibit a non-linear response in the output signal, resulting in an out-of-phase stress–strain relationship. Therefore, in addition to the fundamental, harmonics can be detected over a wide range of frequencies.

The MS. of this paper was received at the Institution on 4th November 1969 and accepted for publication on 14th November 1969. 34
* *Research Student, Mechanical Engineering Department, University of Salford, Peru Street, Salford 3, Lancs.*
† *Lecturer, Mechanical Engineering Department, University of Salford, Peru Street, Salford 3, Lancs.*

In an attempt to understand how these harmonics or non-linearity arise, a linear carrier fluid was investigated in which the percentage concentration of solid suspension was varied from a very dilute dispersion to one of colloidal nature. The type of solid suspension used in this investigation is an oil-furnace carbon black. It is common practice in certain industrial applications to use carbon blacks as filler or strengthening agents in various polymer systems. However, in this work carbon black was used as a base for a suitable high-temperature grease.

A major deficiency of soap-base greases is the melting of the gelling agent (e.g. lithium) at temperatures well below 250°C. It was hoped that by using a non-melting base in the formulation of a grease, very much higher operating temperatures could be attained. Although only one type of carbon black is reported in this work, tests are currently being carried out on other carbon blacks of smaller particle size, over a wide range of temperatures, strain amplitudes, and frequencies. A complete cover of the frequency and amplitude dependence of the system should therefore be provided.

Apart from providing useful information concerning the formulation of a new high-temperature grease, the type of system reported in this work should also show the action of dispersed systems under oscillatory motion conditions.

SUSPENSION

As mentioned previously, the type of suspension used in this investigation is a super-conductivity oil-furnace carbon black, of brand name Vulcan XXX. Furnace and channel blacks differentiate themselves from the commercial carbon because they are of colloidal dimensions. There are a number of grades of carbon blacks, ranging in size from 4000 Å down to approximately 100 Å in diameter. Consequently, the surface areas of carbon blacks differ, and will depend largely upon the definition used. The N_2 absorption reading will probably provide more satisfactory results.

The analytical and particle size data for Vulcan XXX are listed in Table 8.1. Fig. 8.1 shows the frequency distribution for this black. The average particle diameter of Vulcan XXX is 162 Å; but Fig. 8.1 shows that this particular black has a proportion of coarse particles in the 450–600 Å range. The MS 550 fluid used in this work had a nominal viscosity of 0·5 poise.

DESCRIPTION OF EXPERIMENTAL SYSTEM

The system used in this investigation consisted of a basic Weissenberg rheogoniometer manufactured in this country by Sangamo Controls Limited. The shearing geometry is of the conicylindrical type with 0·025-in radial gap and angle of 1° 26'. This type of geometry was first reported by Mooney and Ewart (1)*, and it is usually referred to as the Mooney geometry. When high rates of shear are required, ring and plate geometry can be used.

* *References are given in Appendix 8.1.*

Table 8.1. Analytical and particle size data for Vulcan XXX

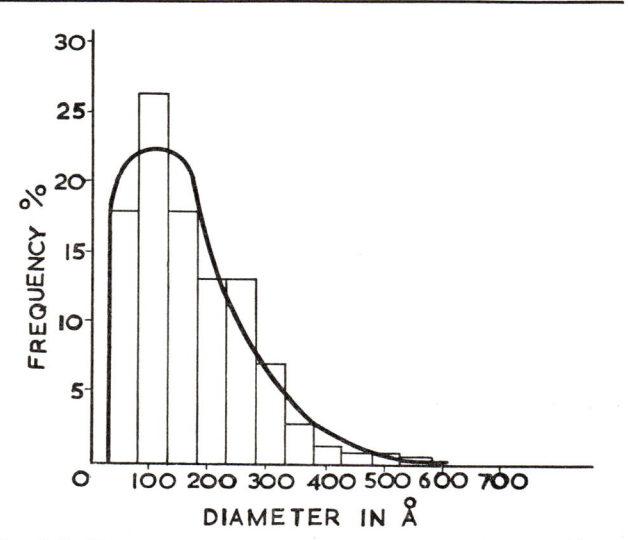

Nigrometer scale	84
Volatile content	1 per cent
Approximate density	20 lb/ft³
pH	9·5
Arithmetic mean diameter	162 Å
Surface average diameter	270 Å
Surface area from size distribution	119 m²/g
Surface area from N_2 absorption	220 m²/g

Fig. 8.1. Frequency versus particle diameter distribution curve

The method used to analyse the output signal from the material under test is that reported by Bogie and Harris (2), which permitted a more precise analysis of the results, especially at the lower concentration.

DISCUSSION AND ANALYSIS OF RESULTS

Viscosity dependence, the dependence of viscosity on the concentration, is shown in Fig. 8.2. Non-Newtonian characteristics were obtained for all concentrations and ranges of shear rate. One of the most interesting observations regarding concentration is that even for very dilute systems, i.e. 1–2 per cent, non-Newtonian behaviour is obtained, this being more pronounced over the lower range of shear rates (10^{-1}–10 s^{-1}). For these corresponding low concentrations, as the rate of shear is increased above 10 s^{-1}, a region is being approached where the viscosity is approximately constant, and a Newtonian régime is presumably being approached.

The question of the settlement of the lightly dispersed systems is illustrated by the 1 per cent concentration in Fig. 8.2, which had previously been sheared and then allowed to stand for 3 h before retesting. Very nearly identical observations were obtained; suggesting that little if any settlement had occurred over this time.

Increasing the concentration produced a more viscous system, resulting in much higher viscosities being obtained at the lower and higher rates of shear. Fig. 8.2 indicates how the rate of breakdown and particle alignment of the structure is influenced by concentration. It is suggested that at the lower concentrations, particle alignment is the predominant factor; alignment and breakdown being of

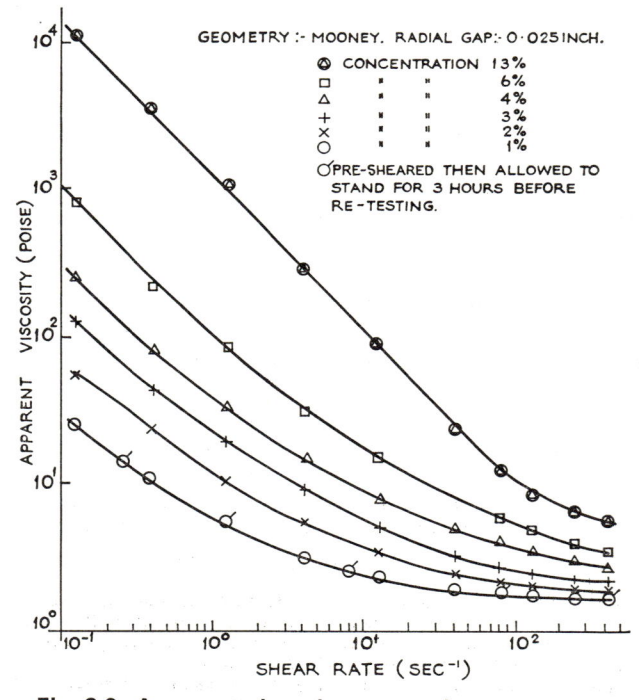

Fig. 8.2. Apparent viscosity versus shear rate curves

secondary importance. However, at the higher concentrations (13 per cent), alignment and particle breakdown occur. These conditions are approximately constant over the range of shear rates 10^{-1}–10^{2} s^{-1}, producing a viscosity–shear rate curve of constant slope over very nearly the whole range of shear.

AMPLITUDE DEPENDENCE

The stress amplitudes of the fundamental frequency (Fig. 8.3) suggest that two flow mechanisms occur over the range of strain amplitudes investigated. For systems with a solids concentration of 3 per cent and over, a range of strain amplitudes covering 10^{-4}–2×10^{-2}, the fundamental stress amplitudes are approximately constant. This would suggest that over this range of strain amplitude little, if any, bulk movement of the suspension is taking place.

Increasing the strain amplitude above 2×10^{-2} produced quite a noticeable increase in the fundamental stress amplitudes. Presumably, bulk movement of the particles (flocculated groups) occurred, resulting in the joining or build-up of the flocculated groups of particles, together with the increase in Brownian motion associated with the increase in strain amplitude.

Further increases in the strain amplitude produce results similar to those observed at low-strain amplitudes, in that the plateau regions in the stress curves are being attained. One possible explanation for these sensibly constant regions of stress could be due to saturation of the system in terms of particle breakdown and alignment.

Fig. 8.3 also shows that there is no direct relationship between concentration and magnitude of the fundamental.

For a sixfold increase in concentration (i.e. from 1 to 6 per cent), an approximate fivefold increase in the maximum stress level is produced. On increasing the concentration from 6 to 13 per cent, a tenfold increase in the fundamental stress amplitude is produced. These two observations illustrate the differences in stress amplitudes between dilute and concentrated colloidal systems.

It is of interest to observe that even for very dilute systems the response to sinusoidal shearing motion is non-linear. The non-linear response suggests that the inertia of the particles in dilute systems is sufficient to cause a disturbance to the bulk flow of the carrier fluid (MS 550). Fig. 8.4 illustrates how the third harmonic stress amplitude varies with strain amplitude for the various concentrations. For the lower concentration (2 per cent) the magnitude of the third harmonic stress amplitude is quite small, in the region 0–10^{-1} in strain, but increases quite noticeably as the strain amplitude is increased beyond 10^{-1}.

The fifth harmonic stress amplitude (Fig. 8.5) is non-existent for the lowest concentrations, and is just detectable for the 2 per cent concentration for strain amplitudes 10^{-1}. The general indication of the graphs suggests a positive increase in value and agrees with the third harmonic stress amplitude observations. As the concentration of solid particles is small compared with the volume of carrier fluid, it is suggested that this non-linearity must be due to inertia of the particles, as collision and breakdown of particles must be of secondary importance for such concentrations.

Increasing the concentration produces an increase in the magnitude of the harmonic stress levels. Over the initial range of strain amplitudes the third harmonic stress amplitudes decay slightly for all concentrations, and then increase quite sharply as the strain amplitude increases.

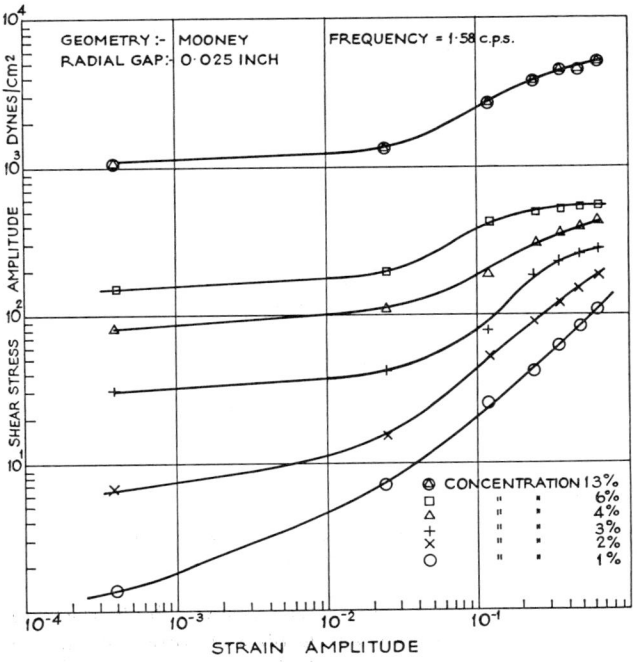

Fig. 8.3. Fundamental stress amplitude versus strain amplitude curves

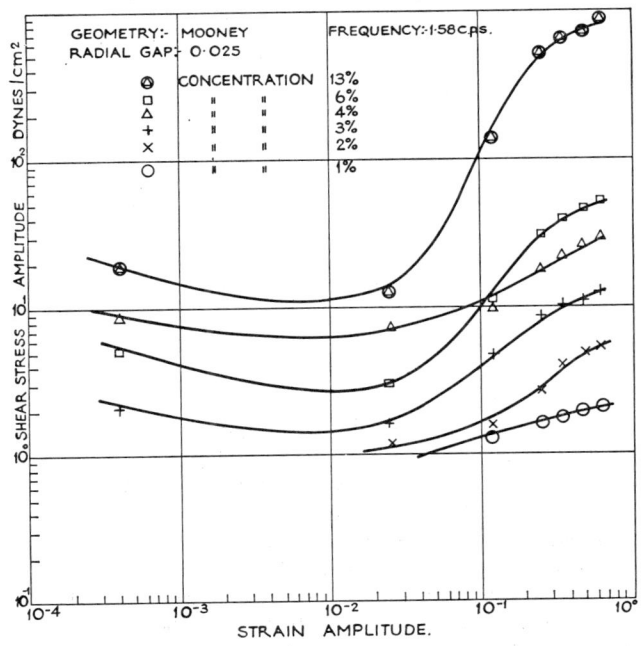

Fig. 8.4. Third harmonic stress amplitude versus strain amplitude

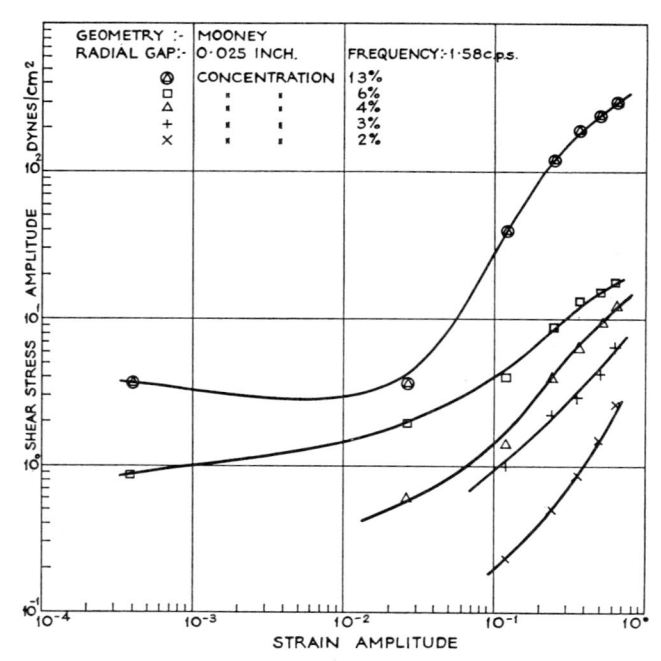

Fig. 8.5. Fifth harmonic stress amplitudes versus strain amplitudes

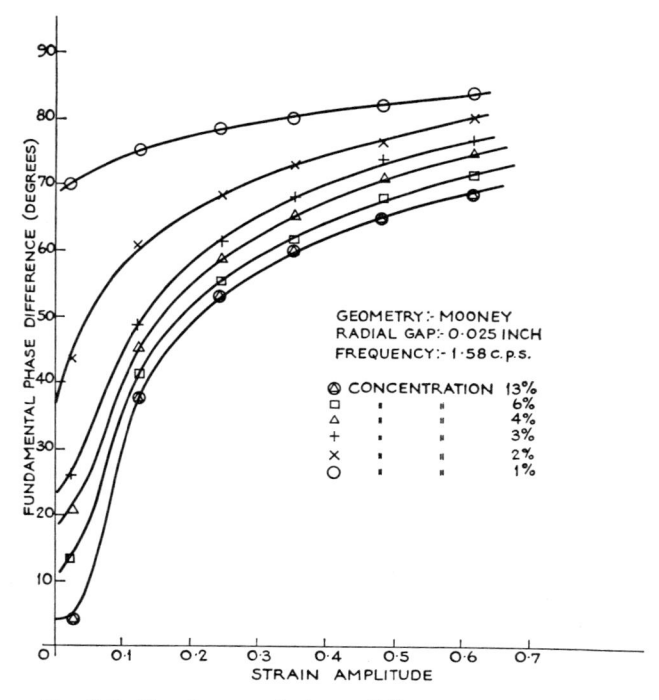

Fig. 8.6. Fundamental phase difference versus strain amplitude

Figs 8.4 and 8.5 suggest that the non-linear content over a range of strain amplitudes $0-3 \times 10^{-2}$ remains approximately constant, suggesting an independence of strain amplitude on the grease. However, as the strain amplitude is increased the non-linear content increases, and the system becomes more amplitude dependent.

When discussing the fundamental stress amplitudes under 'Amplitude dependence' (Fig. 8.3), it was suggested that two regions are present in these curves where the system is observed to flow at an approximately constant level of stress. Fig. 8.6 shows that the fundamental phase differences for a concentration of 3 per cent and for strain amplitudes below 2×10^{-2} suggest elastic-type response, because the fundamental phase differences are small. Between strain amplitudes 3×10^{-2} and 4×10^{-1} the phase difference, irrespective of concentration, increases quite rapidly, the response being of a more viscous nature.

When the plateau regions are approached (strain amplitudes 5×10^{-1}), the fundamental phase differences are of a viscous nature. The region in which this viscous response is observed coincides with the region of strain amplitudes in which the fundamental stress amplitudes are approximating to plateau regions. The previous suggestions regarding break-up and alignment appear to be correct, the systems now acting as very dilute dispersions. The predominant response is that of the carrier fluid, which for the given set-up produced a complete viscous response at all strain amplitudes.

The harmonic phase differences (Figs 8.7 and 8.8) indicate trends with concentration and strain amplitude. In almost all cases, increasing the strain amplitude from the minimum value to a strain amplitude of 2×10^{-3}

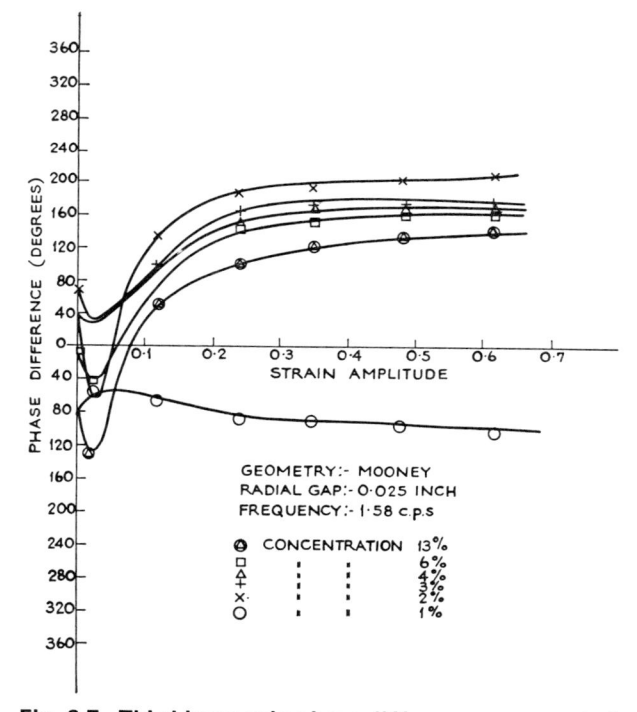

Fig. 8.7. Third harmonic phase difference versus strain amplitude

produces harmonic phase differences which decrease in value. In most cases a negative phase is produced which decreases negatively and finally becomes positive. These curves suggest that the phase shift for each concentration

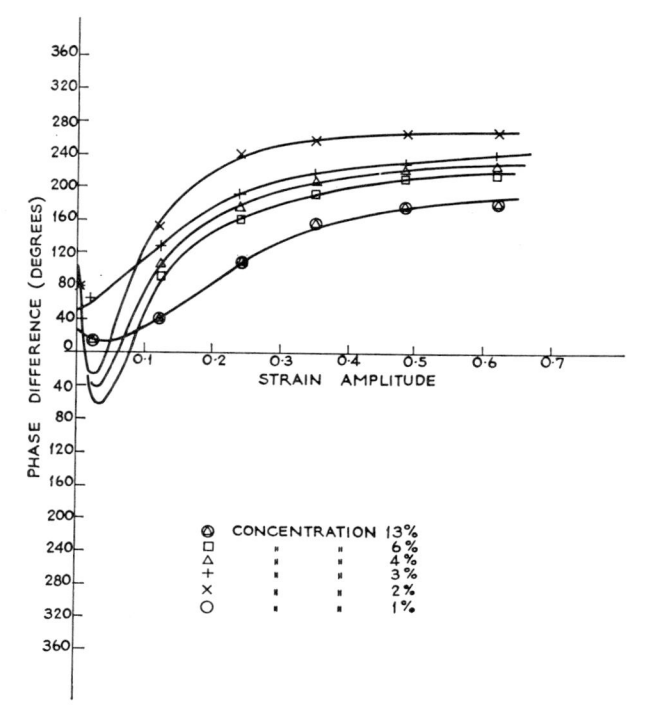

Fig. 8.8. Fifth harmonic phase difference versus strain amplitude

becomes constant for strain amplitudes greater than 2×10^{-1}. These observations are of interest because Bogie (3) checked similar systems for their frequency dependence and reported quite large harmonic phase shifts.

IMPROVEMENT OF THE MECHANICAL AND TEMPERATURE STABILITY

The use of carbon blacks as thickening agents in synthetic greases in their present form does not hold much promise of useful results. The mechanical stability of such systems to shear is very poor, but their temperature stability is superior to many soap greases. To improve both the shear and the temperature stability of the system described here, a structure chemically similar to that formed in thermally formed greases is needed. In such a system the soap fibres, plates, etc., are formed *in situ*, and most of them possess a lyophilic nature, a feature associated with mechanical stability.

Synthetic thickened greases do not possess this lyophilic characteristic. Therefore, to improve the shear and temperature stability of the system being considered, a synthetic product with properties similar to those of soap grease was required. Finally, a method which treated particles prior to mechanical formulation of the grease was studied. The carbon black particles were treated with a solution of methyl-T-S, using a solvent method. It is regretted that this cannot be divulged. Treatment of the surface of each carbon black particle varied from 0·2 to 5 per cent by weight; the worked consistency, bleed rate, and volatility content were checked. From this initial work it was determined that the optimum percentage treatment for the carbon black investigated here was 2 per cent by weight.

In order to ascertain whether a lyophilic nature similar to that inherent in thermally formulated greases had been attained, grease samples were prepared, using the same percentage by weight concentration given previously. These were tested in continuous shear in the range of $10-4 \times 10^4$ s^{-1}, using ring and plate geometry. This rate of shear was thought to be realistic and as encountered in normal practice. Identical temperature ranges were used for each test; the results of one of these tests are shown in Fig. 8.9. The variation of apparent viscosity with shear rate is presented for three different greases. The synthetic grease with 7 per cent concentration of treated carbon

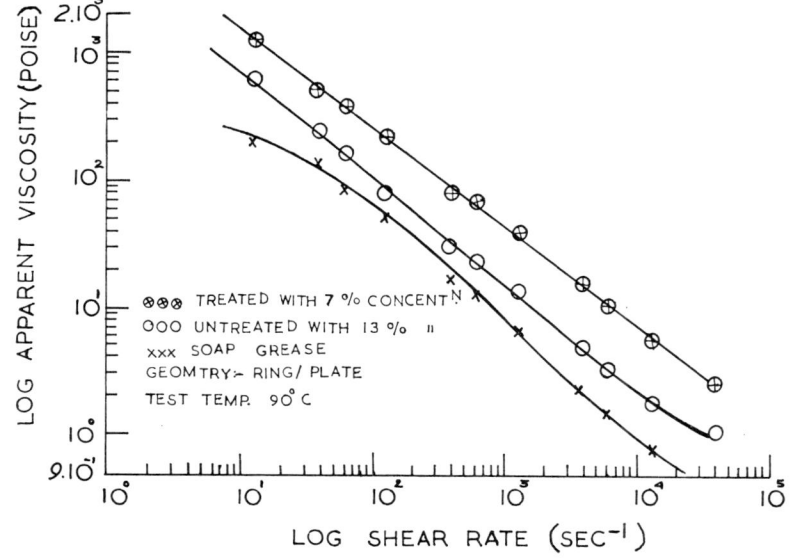

Fig. 8.9. Apparent viscosity versus shear rate

blacks had higher viscosities than that with 13 per cent concentration of untreated carbon blacks for the lower and higher shear rates. In addition, it was determined that the treated synthetic grease is superior to that of a soap grease having the same carrier fluid. In the former no decomposition results with temperature, in contrast to most soap greases where decomposition occurs when the operating temperature is raised.

Treatment of carbon black particles resulted in a grease of higher consistency for lower concentration of solid (7 per cent), which has a penetration value of 226 (NLGI grade 3).

This showed that a system (lyophilic) similar to that present in soap greases had been produced and that similar synthetic solids can be treated to form stable systems.

CONCLUSIONS

In conclusion, it would appear from the work reported in this paper that suspensions formed from solid concentrations exhibit non-Newtonian characteristics even at very dilute solids concentrations (1 per cent). Furthermore, it has been shown that the rheological behaviour of such a system cannot be treated as a linear viscoelastic system. In addition to the fundamental phase and stress amplitude, harmonics of the fundamental are present. While the harmonic phase differences have been shown to be independent of the strain amplitude, previous work has shown them to be frequency dependent.

The mechanical shear stability and consistency of the above system were improved by special surface treatment of the carbon blacks. The optimum percentage treatment of these blacks, using a special solvent method, was found to be 2 per cent by weight. Work in this field is currently being carried out to investigate the possibility of producing a synthetic grease with larger size solid particle in order to produce a less expensive grease which is stable to high temperature and mechanical shearing.

ACKNOWLEDGEMENT

The authors would like to thank Professor J. Halling for making it possible to carry out research in this field, and the Esso Petroleum Company Limited for supplying the samples.

APPENDIX 8.1

REFERENCES

(1) MOONEY, M. and EWART, R. H. *Physics* 1934, **5**.
(2) BOGIE, K. D. and HARRIS, J. 'An experimental analysis of non-linear waves in mechanical systems', *Rheol. Acta* 1967 (6th January)
(3) BOGIE, K. D. 'The non-linear response of soap and synthetic greases', *Trans. Am. Soc. mech. Engrs* 1968 (July).

Paper 9

SOME FAILURES OF GREASE-LUBRICATED ROLLING-ELEMENT BEARINGS

E. D. Yardley*   W. J. J. Crump†

INTRODUCTION

THE NAME 'rolling-element bearing' covers a wide variation in design and construction. One feature which is common to the majority of applications is the repeated stressing of the rolling elements and races. Under ideal conditions of load, speed, and lubrication, and in the absence of any gross faults in the materials, this repeated stressing will eventually cause the bearing to fail in fatigue. Such a failure manifests itself in the form of a flake or flakes of material detached from the running track of one of the races or possibly the rolling elements (Fig. 9.1). As the production of a fatigue flake is dependent upon a suitable stress-raiser being present, which is a random occurrence, it follows that the fatigue process is random, and that there will be a scatter in the lives obtained from bearings run under identical conditions. This scatter follows the Weibull distribution, and thus some bearings are bound to fail in a comparatively short time. It is important to be aware of this when bearing failures are being examined, for if a bearing has failed in a time shorter than the life given by the manufacturer's recommendations, it may simply have been at the bottom end of the distribution.

The manufacturer's ratings give the life which will be exceeded by 90 per cent of the bearings. Those which achieve this rarely elicit comment, in contrast to those which do not. It is important, therefore, that these early fatigue failures are recognized and that they are not blamed on other causes. However, most early failures have not been caused by fatigue but are due to some form of ill-treatment.

Factors which can precipitate bearing failure are many and varied, but fall generally under the following headings:

The MS. of this paper was received at the Institution on 29th September 1969 and accepted for publication on 29th October 1969. 33
* *National Coal Board, Mining Research and Development Establishment, Ashby Road, Stanhope, Bretby, Burton-on-Trent, Staffs; formerly G.E.C. Power Engineering Ltd, Leicester.*
† *Mechanical Engineering Laboratories, G.E.C. Power Engineering Ltd, Whetstone, Leicester LE8 3LH.*

(1) Material faults, e.g. incorrect heat treatment, slag inclusions, etc.
(2) Failure of associated components, e.g. failure of adjacent thrust bearing, leading to excessive load on lipped roller bearing.
(3) Incorrect application; this includes incorrect estimate of service conditions.
(4) Human agencies, e.g. bad fitting.
(5) Lubricant, e.g. unsuitable grease.

Although any of the factors included in the above list can by themselves cause failure, it is often found that it is a combination of factors which is responsible for shortening the bearing life. With the continual demand in industry for greater speeds, higher loads, and more elevated temperatures there is a growing tendency for bearings to be run at or near their limits. Although, in general, these limits include a safety margin, bearings run under these conditions have little capacity to absorb adverse changes in circumstances. Thus a small increase in the severity of running conditions, which would be tolerated in less trying circumstances, can precipitate failure.

It is the purpose of this paper to consider bearing failures for which the grease, either by itself or in combination with other factors, was responsible or thought to be responsible. Grease may be at fault due either to the nature of the grease itself or to its application.

Taking first the nature of the greases, three ways can be found in which unsuitability may arise. These are first, the grease is simply not a rolling-element bearing grease; second, it is not suitable for the temperature; and third, it lacks resistance to solvents.

The first of these three points is well illustrated by the example of a chassis grease which is fibrous and sticky, and will not clear from a bearing. As a result, the temperature in the bearing, from starting up, rises until failure occurs either by the grease going hard and ceasing to lubricate or the bearing clearance being taken up.

Unsuitability with respect to temperature generally involves elevated temperatures, although greases can be

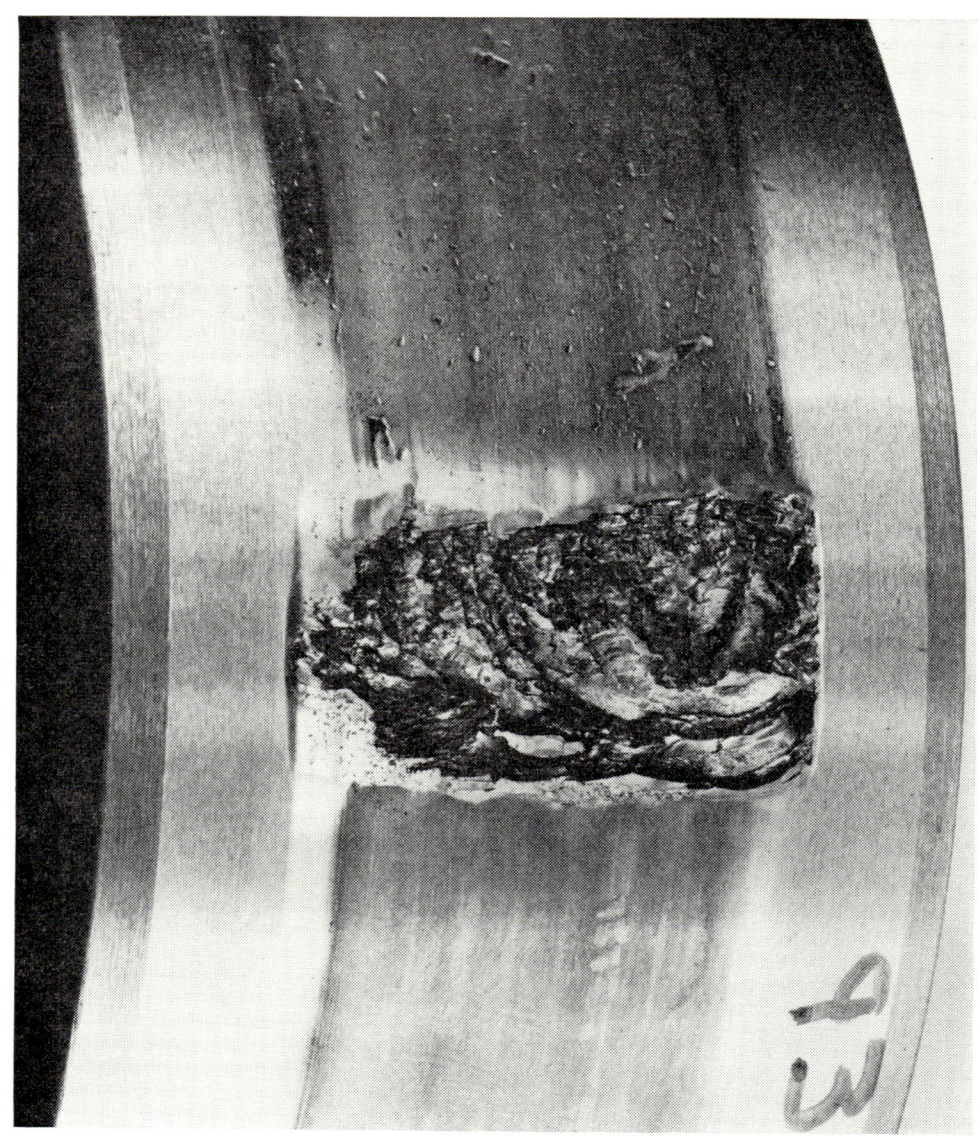

Fig. 9.1. Typical fatigue flake

unsuitable due to their cold torque characteristics. At elevated temperatures a grease with too low a drop point will melt and run out of the bearing. Alternatively, greases can go hard and leathery and fail to lubricate.

The lack of resistance to solvents generally refers to water washout characteristics and the ability of the grease to prevent rusting of a bearing, but sometimes other fluids are involved. Where lubrication is required in such environments, certain special greases are produced to be resistant to particular fluids, such as petrol.

The term 'application' as used above encompasses all aspects of the provision of grease to a bearing, including the initial packing, relubrication, and problems associated with the relubrication process.

The principal faults which are encountered in the initial packing of a bearing are overpacking and contamination. Overpacking does not leave room for the grease to clear from the bearing. The temperature rise which is observed when a bearing is first started, and which is due to churning during the period in which the grease is clearing from the bearing, continues in an overpacked bearing, and failure occurs when the grease dries out and ceases to lubricate. The ingress of dirt during packing may result in debris pitting, with possible resultant fatigue flaking or severe wear of cages or rolling elements.

In certain of the more arduous applications it is vital that relubrication is performed at fairly frequent intervals, particularly when bearings are running close to their rated maxima. Failure to relubricate is one of the factors which, as stated previously, might be tolerated in less severe circumstances, but could contribute to premature failure in arduous conditions.

During relubrication there is the possibility that dirt may get into the bearing through the grease pipes, as these have a tendency to collect a plug of dirt in their ends which is pushed into the bearing in front of the fresh grease. The fresh grease itself may be a source of contamination, as it is not unknown for bearings to be relubricated with grease which is unsuitable for the purpose or with dirty grease. In addition, there remains the possibility that the grease pipes are blocked, perhaps due to hardening of the grease, thus preventing relubrication. Contamination of bearing grease can also occasionally arise from lubricants used in adjacent machinery, such as gears.

This somewhat brief discussion indicates some of the ways in which grease may be involved in bearing failures. The sections below describe some examples of failures of rolling bearings in which the performance of the grease was an important factor.

BEARING FAILURES
Grease contamination

In general, when failures of traction motor bearings occur they do so at some situation inconvenient for examination, and the resultant removal of the machine to a suitable place for stripping reduces the bearing to several handfuls of unrecognizable steel scrap. Occasionally, however, a failure is caught at an early stage in its development, and this was the case when a 70-mm bore commutator-end bearing was returned for examination. The bearing was a cylindrical roller having two lips on the outer race, and one lip on the inner, with a retaining flange completing the four-lip assembly (Fig. 9.2). The cage was of machined yellow metal. Together with the bearing, samples of grease taken from the commutator-end pipe, the pinion-end pipe, and the grease recess on the inner side of the bearings were received.

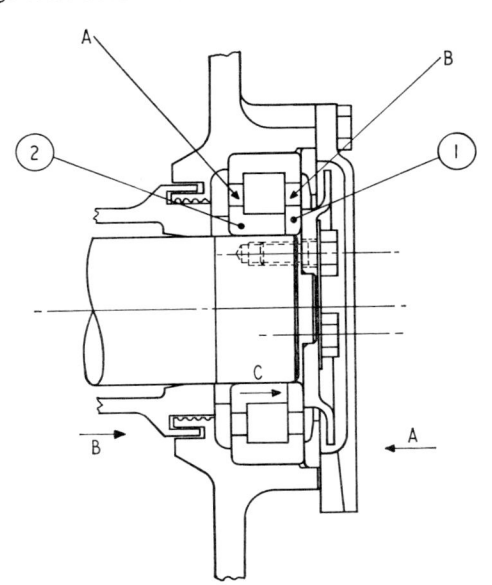

Fig. 9.2. Commutator-end bearing showing four-lip arrangement

A visual examination of the inner race retaining flange showed signs of fretting in the regions of the end cap securing bolts, and a worn rim on the inside where contact had been made with the rollers. The latter showed a corresponding rim on their outer ends (Fig. 9.3), but overheating and smearing were evident on their inner ends. The lip on the inner race, which had been adjacent to the overheated ends of the rollers, also exhibited evidence of overheating and oxidation (Fig. 9.4), as did the lip on the outer side of the outer race.

Microhardness tests on the rollers showed that the overheated areas were considerably softer (590 V.p.n. against 868 V.p.n.) than the rest of the rollers, while the worn rim was not. On the retaining flange the worn region proved to be slightly harder (792 V.p.n.) than the rest of the material (753 V.p.n.).

Before it had been dismantled the bearing had been washed in white spirit, and some of the material from inside the bearing was taken for spectrographic analysis. The results indicated that the material consisted of wear debris from the bearing and cage, as expected, and dirt from the environment.

From this examination it was concluded that the condition of the bearing was due to either a fairly large axial load together with an axial impact, or a fairly large axial load which reversed from time to time. In either case a deficiency of lubrication existed at the roller ends, and the ingress of dirt to the outer side of the bearing resulted in the wear rims on the retaining flange and the roller ends.

In this instance, then, several factors combined to cause bearing failure, which, although not well advanced, could quickly have led to more serious damage being done in the bearing.

Unsuitable grease

A number of bearing failures have arisen from advanced development work on machines. The ancillary equipment is used under conditions which do not quite conform to the original specifications.

Such a case was encountered when a number of failures of a double-row barrelled roller bearing of 1·4 in internal diameter were reported. The bearings were used in a fuel pump which was intended to pump a fluid of very low viscosity and poor lubricating characteristics. As the fluid was not suitable for lubricating the bearing, a premium grade 3 lithium grease was used. Examination of two failed bearings revealed gross cage wear (Fig. 9.5), wear of the running tracks and one end of the rollers, and some evidence of corrosion. It seemed highly likely that the fluid being pumped was finding ingress into the bearing, was not compatible with the grease, and was washing the latter out of the bearing. Subsequently, the observed wear of the cage and the races occurred. As a check, a washout test on this grease was carried out using the fluid being pumped. The results confirmed that the grease was completely destroyed by the fluid in a short time. At the time of writing, a successful conclusion to this problem

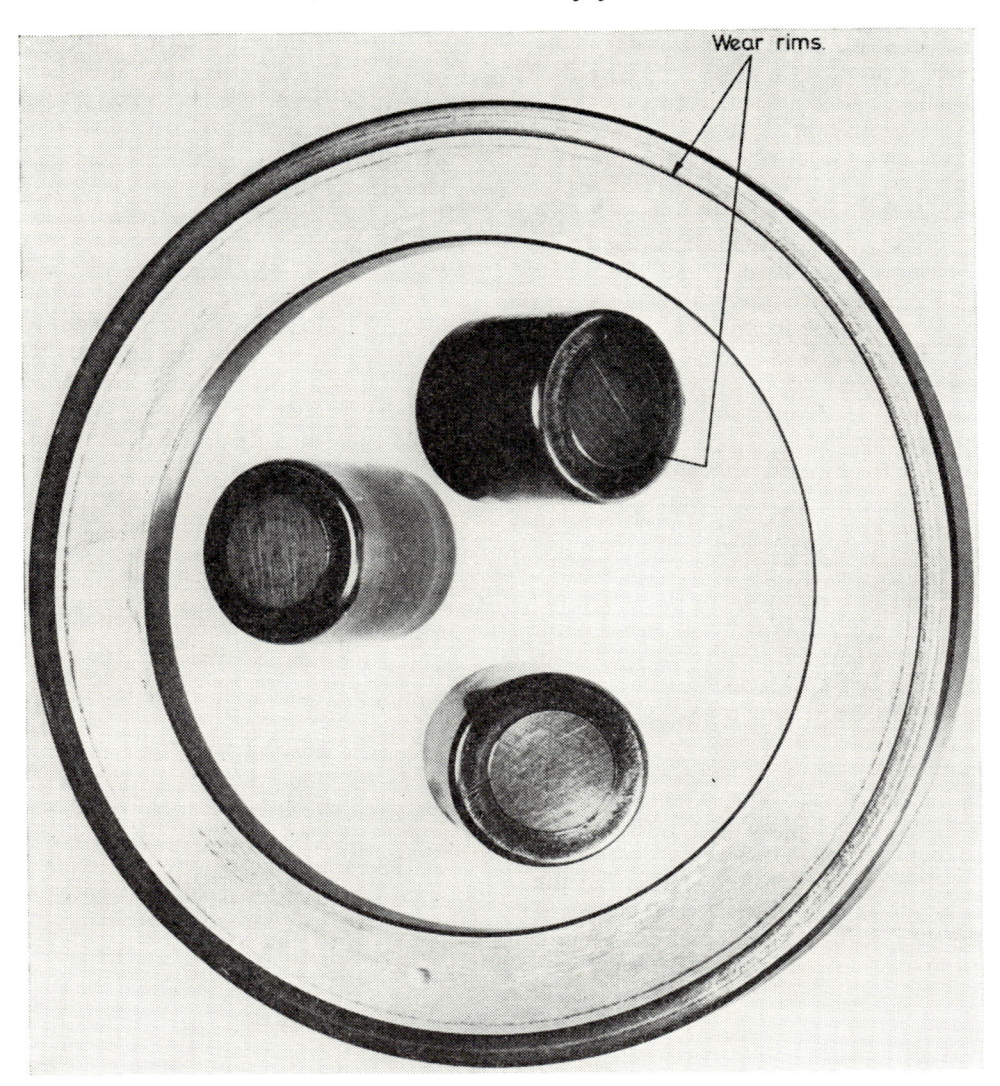

Fig. 9.3. Retaining flange and rollers showing abrasive wear

had not been reached, and work was in hand to try to prevent the fluid from entering the bearing without major modifications to the pump design and to find a grease with suitable washout characteristics.

Overheating

Failures of bearings due to overheating are in many cases attributable to lubrication failure arising from bad packing or using the wrong grease. Thus, when an externally aligning 110-mm bore roller bearing was returned from service in an obviously overheated condition, the grease was immediately suspected as being responsible for the failure. The bearing, which had been used as a generator commutator-end bearing, was in very poor condition when received, had lost its self-aligning properties, and would no longer function as a bearing. The steel was discoloured and suggested that temperatures in the region of 300°C had been reached, and the grease had been reduced to a carbonaceous deposit in the bearing. However, when the inner race was removed it was found to be bright and shining, with no visible evidence of overheating. Prominent gouge marks were found running across the bearing surface of this race, and severe pitting had occurred in the proximities of these marks. Similar marks had been previously observed on failed bearings of the same type and were found to be due to misalignment during fitting. The inner race had been sleeved to take a shaft of 102 mm diameter. The inside of the sleeve was fairly heavily pitted in a band around its circumference with two cracks which extended for the full thickness of the sleeve, one covering approximately half and the other a quarter of the length of the sleeve. Thus, at this stage, the bearing failure could have been ascribed to three factors: misalignment during fitting, cracks in the sleeve causing fretting, and failure of the grease to lubricate, although the last of these was difficult to reconcile with the condition of the inner race.

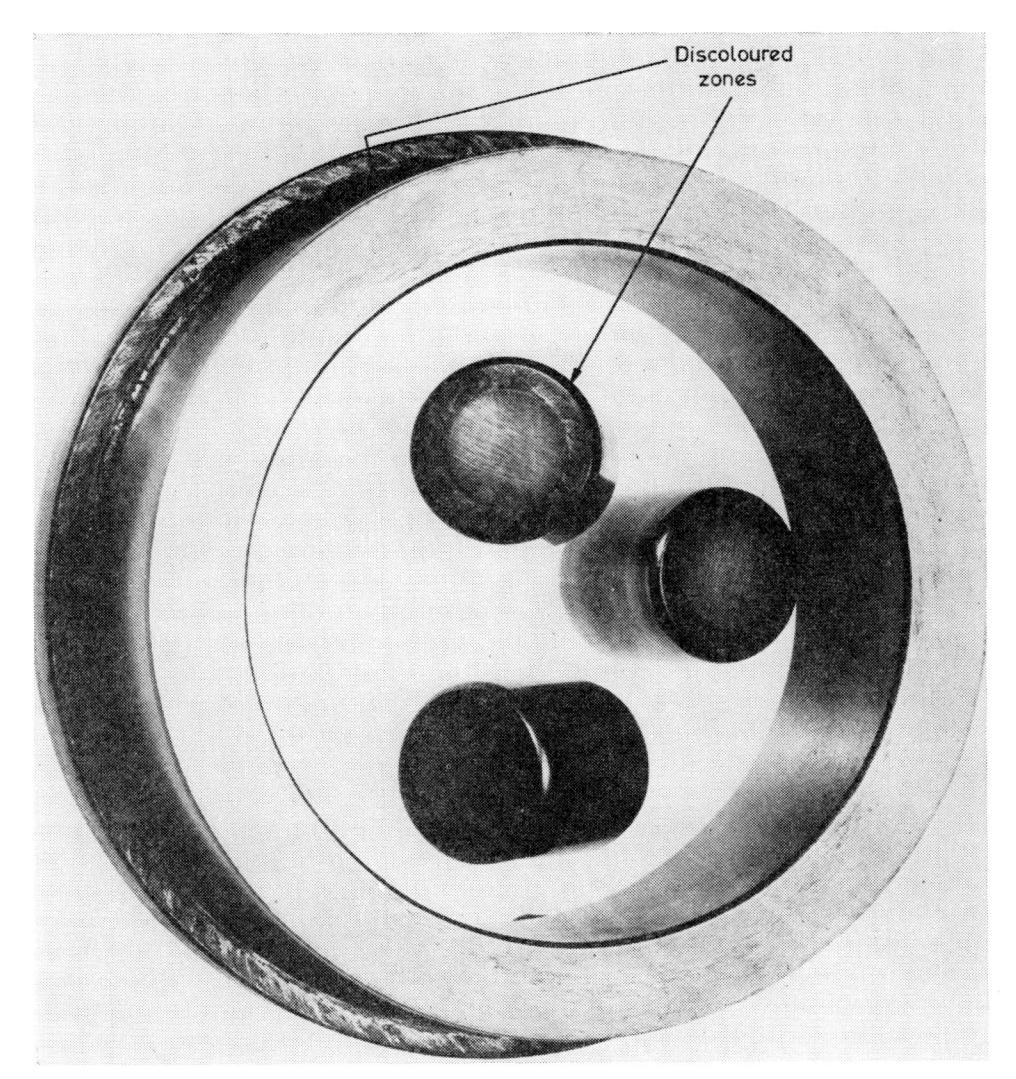

Fig. 9.4. Inner race flange and rollers showing overheating

The cage was dismantled and removed, together with the rollers, in order that the running track of the outer race could be examined.

The indications of overheating shown by the external appearance of the outer race were again present on its bearing surface. The discoloration had a mottled appearance with straw-coloured and blued patches.

On approximately two-thirds of the circumference the mottled discoloration was broken up by equi-pitched triangular straw-coloured areas (Fig. 9.6). The mottled effect on the remainder of the track showed a predominance of the blued component, and some surface deterioration was present in this region (Fig. 9.7).

In general, the bearing surfaces of the rollers showed heavy discoloration. Twelve of the 18 rollers were lighter coloured on one side than the other, and on the darker side of some of these a triangular area was outlined by carbonaceous deposits. On the ends of the rollers, areas were found corresponding to the position of the cage. On some rollers, areas not covered by the cage were covered with the carbonaceous deposit, which was presumably the residue from the burnt grease. Examination of the cage revealed little except the deposit resulting from the heating of the grease.

At this stage, therefore, the reasons why the bearing failed were somewhat confused. Clearly, misalignment during fitting had been the cause of the gouge marks on the inner race, but was unlikely to be responsible for the overheated condition of the bearing. The bright condition of the inner race conflicted with the obvious overheating of the outer race, and it seemed unlikely that a lubrication failure was responsible for the high temperature. The presence of the triangular areas on the outer race and the rollers suggested that the bearing had been heated from some external source while stationary, and the grease between the rollers and the race had protected parts of

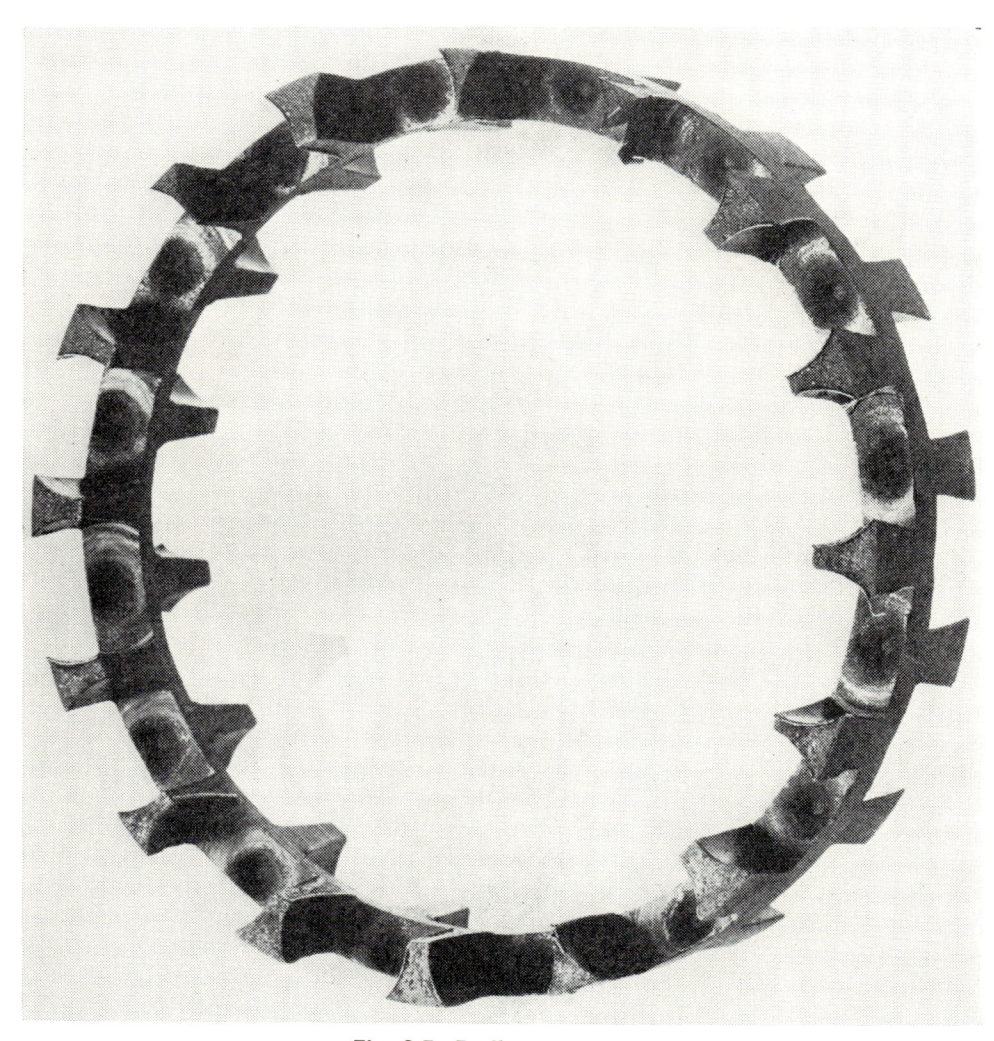

Fig. 9.5. Badly worn cage

both at the points of contact. In order to test whether this hypothesis was feasible, a fully greased 12-mm roller bearing was heated at 245°C for 2 h under a small load. Inspection of this bearing after the test revealed that parallel-sided markings had been formed on both races in positions corresponding to those of the rollers. The latter were marked in a similar fashion to those in the large bearing with protection due to the cage on their ends in places. The condition of the grease was slightly better than that in the large roller journal, but most of the grease had run out of the bearing. This test indicated that stationary heating was possible as a cause of failure. In order to throw some light on the possible source of such heating, hardness measurements were made on the 120-mm bearing and compared with those from an identical bearing which had suffered no heating. The results showed that the outer race was somewhat softer than the inner (695 V.p.n. against 761 V.p.n.), and that both were softer than those from an unheated bearing.

These results suggested a heating from around the outside of the bearing, as the hardnesses were fairly uniform around the circumference of the races. Such a heating would not provide uniform temperatures on inner and outer races, in contrast with small bearings heated in the furnace. It was noticeable that markings were formed on both races of the small bearing but only on the outer of the large bearing. As the inner race of the latter had undergone some softening it was assumed that it must have been *in situ* during the heating, and it was concluded that due to its distance from the heating source, and the fact that the shaft would form an efficient sink, the temperature reached by the inner race was not sufficient to cause discoloration. From the condition of the outer race it appears that the bearing did not run after the overheating took place.

With reference to the condition of the sleeve on the inner race, it was assumed that some movement between the sleeve and the shaft caused the fretting and cracking observed.

Thus it was concluded that the failure, which at first

Fig. 9.6. Outer track showing discoloration in a regular pattern

Fig. 9.7. Outer track showing overall discoloration

sight appeared due to lubrication failure, involved two separate causes:

(1) Gouge marks on the inner race were caused by misalignment during fitting, and the pitting associated with these would have produced fatigue flakes on subsequent running. This condition would by itself constitute a bearing failure.

(2) External heating, more or less uniformly round the outside of the bearing while this was stationary, dried out the grease, caused oxidation of the outer race and parts of the rollers, and resulted in the self-aligning properties of the bearing being lost. Again, this condition by itself is a failure condition. While no record of any heating having been applied to the bearing could be found from its history, the possibility remains that some unauthorized attempt at stripping caused the heating concerned. However, beyond this, the source of the heating remains the subject of speculation.

Thus, in this instance, there are two failures in one, neither being one which would be deduced from a cursory examination of the bearing.

Lubricant deterioration

This example well illustrates the statements made previously regarding the multiplicity of factors which can combine to cause bearing failure.

A lipped cylindrical roller bearing was removed from the non-drive end of a 400-hp, 3000 rev/min electric motor during a routine inspection. The bearing was a four-lipped type with an angle ring providing the fourth lip. In correspondence concerning the bearing it was stated that there had been a long history of failures of this type of bearing since the machines were commissioned. The particular bearing in question had failed by a complete disintegration of the cage, parts of which were sent for examination together with a roller and several samples of grease from similar bearings in other motors on the same site. A report on the inspection of the bearing, the grease, and adjacent parts at the time of stripping was sent to provide assistance in determining the cause of failure. The salient points from this report revealed that the motor had had a number of strips and rebuilds in the 90 h or so during which it ran. An outer cover fitted to the non-drive end had rubbed on the shaft causing metal pick-up and fusion. A raised ridge had been formed round the shaft, and the inside face of the cover was burnt and smeared. This fault had cleared itself before the bearing was examined. The space inside the cover was filled with black, badly discoloured grease which should not have been found in this region. There was plenty of grease in the escape chamber of the outer cap but it was also black and discoloured, as was that filling the bearing. On removing the bearing from its housing it had been found that the space in the housing at the point of entry of the grease via the lubricating nipple and pipe did not contain much grease, whereas it would have been expected to be full. The steel cage was in 15 pieces, the 13 spacers separating the rollers having broken from the outside rings (Fig. 9.8). The grease relief slinger showed indications of a rub between the outer cap and the escape face of the slinger.

The parts of the bearing received for examination were the two rings from the cage and six of the spacing bars together with one of the rollers. Fig. 9.9 shows a diagrammatic section through the cage, with labelling of those parts which suffered wear. The projections on the rings, being the remains of the cage bars, were examined, and while some were relatively undamaged others were severely distorted due to wear at the shoulders and rubbing on the fracture faces. Of the 13 projections only four on each ring possessed fracture surfaces suitable for detailed examination. It was possible to identify four cage bars which were broken from one of the rings by the fit of the fracture. The distribution of these cage bars suggested that the fractures had started at two almost diametrally opposite points on the circumference of the rings. On the other ring, three cage bars could be fitted to well-preserved fractures, and another to one of the less severely damaged projections, while the remaining two could not be placed.

The cage bars were examined, and one feature common to all was the scuffing of the flange faces. The wear of the bottoms of the flanges was also similar in each case with a burred-over outside edge on one flange and the bottom of the other flange worn flat. Some rubbing of the fracture surfaces was apparent, but the shoulders of the cage bars showed no wear, in contrast to their remains on the rings.

Examination with a low-power ($\times 8$) magnifier revealed cracks in some of the stubs of the cage bars on both rings and on two of the spacing bars. The cracks in all cases ran in a plane concentric with the circumference of the rings. Certain parts of the fracture surfaces of the stubs and spacing bars were typical of failure in the brittle mode, while others suggested failure in torsional fatigue.

Higher power magnification ($\times 40$) revealed that some cracks had opened considerably and could be seen to be the regions where fractures began. Even on the best preserved fracture faces rubbing had obscured much of the detail.

In one case a crack near to the surface appeared to have spread to the surface and extended progressively through the metal in the other direction. There was evidence that fractures had begun at two points diagonally opposite to each other on the spacing bar.

Following the discovery of these cracks, transverse sections of two cage bars belonging to opposite sides of the cage were taken together with a longitudinal section of one of the bars, and they were prepared for metallographic examination. On each section small slag inclusions lying in strings parallel to the surface of the bars were found. In the longitudinal section, which was taken through a crack in one of the cage bars, a string of slag inclusions ran almost the whole length of the specimen (approximately 4 mm) (Fig. 9.10).

The grain structure of the cage material was severely deformed at the edges where scuffing had taken place, but otherwise was quite regular and equi-axed.

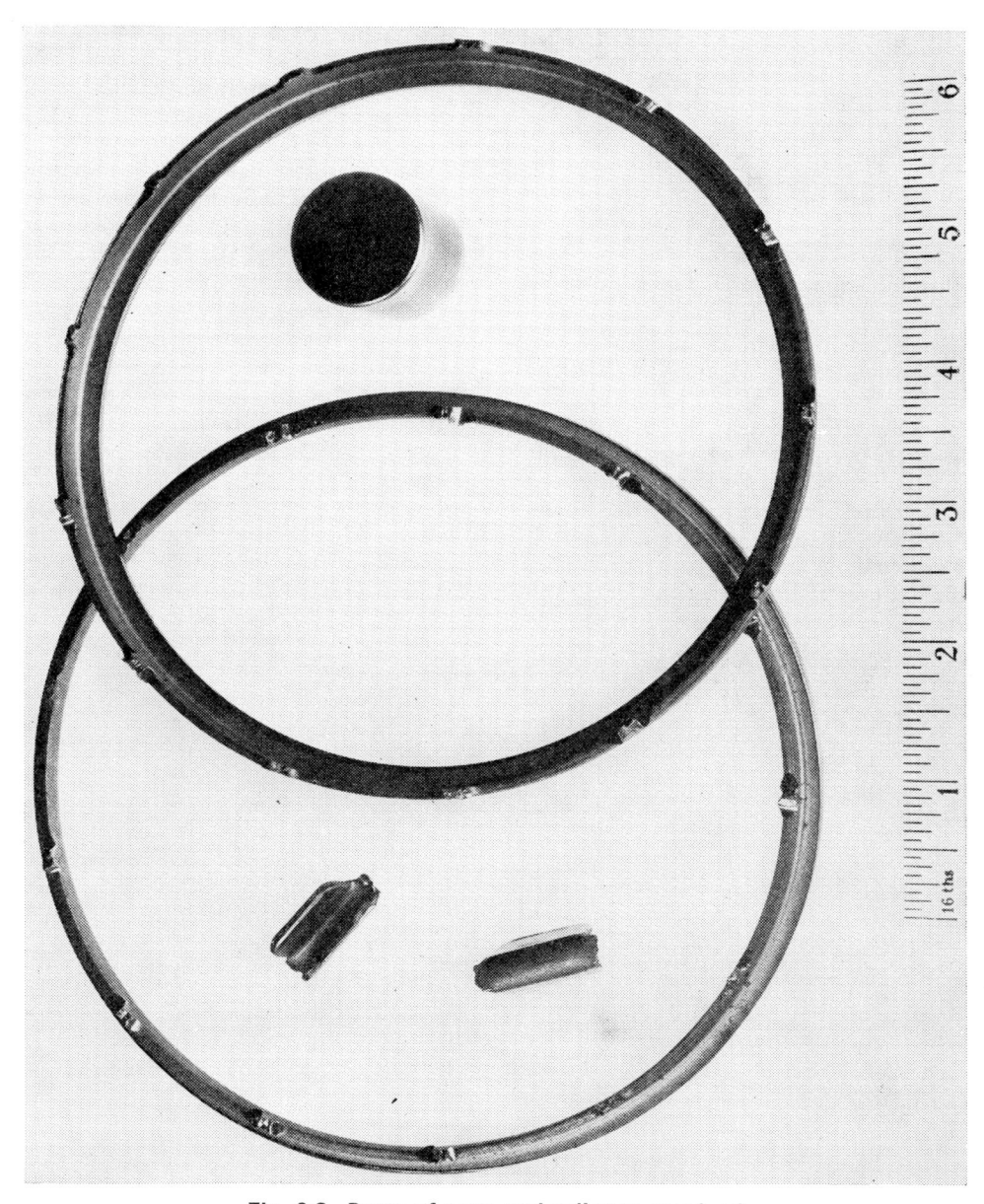

Fig. 9.8. Parts of cage and roller as received

The samples of grease sent with the bearing were analysed spectrographically together with a sample of the grease which was the recommended lubricant for the bearings. Of the seven samples received, two were of similar composition to the recommended grease with the addition of wear debris and some environmental dirt.

Two more appeared to be mixtures of the recommended grease and a calcium grease, while the remaining three samples seemed to be entirely free of the recommended grease and to be composed of a mixture of calcium- and lithium-based greases.

Several important points regarding the reasons for the failure were given in the report on the stripping examination of the bearing. First, the machine had been stripped and rebuilt during the life of the bearing. Second, the outer cover had rubbed against the shaft with sufficient severity to cause pick-up and burning, and the escape face of the grease relief slinger had rubbed against the outer cap. Third, black discoloured grease was found both within and outboard of the bearing. The other points arising from the examination of the material which allowed a sequence of events leading to failure to be put forward included the presence of slag inclusions, scuffing of the flange faces on the spacing bars, the appearance of the fractures, and the grease analyses.

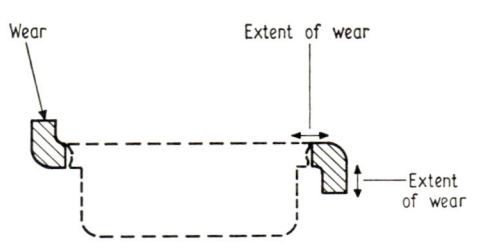

Fig. 9.9. Section through cage, worn surfaces indicated

The progress of the failure was thought to take the following course.

On rebuilding the motor after the strip, the space outside the bearing was, in error, overpacked with grease and the outer cap was not fitted correctly. On starting, the cap rubbed against the shaft causing a temperature rise which, when combined with that due to the grease clearing itself from the bearing, resulted in the lubricating properties being reduced. This lower lubricity led to stick-slip between the rollers and the flanges of the cage spacing bars, thus exerting an intermittent torsional force on the spacers. As the spacers sectioned were those belonging to the best preserved fractures, it might be assumed that these were the last to fail, presumably because the others contained greater flaws, which might, in fact, have been opened into cracks during the process of forming the cage. However, had they not already been opened, their presence as stress raisers on the longitudinal planes would have provided a torsional force of any reasonable magnitude with ample crack nuclei. The first crack thus opened on the longitudinal shear plane of the spacer, and in all probability progressed to the surface via the 45° planes. (Some of the fractures show this type of morphology.) After the first fracture had occurred, the other end of the spacing bar was subjected to a complex system of forces involving both torsion and bending, and this may explain why not all of the fractures were typical of pure torsional failure. When the first spacer had become detached, the removal of the remainder proceeded with increasing rapidity as greater loads were imposed on them. After

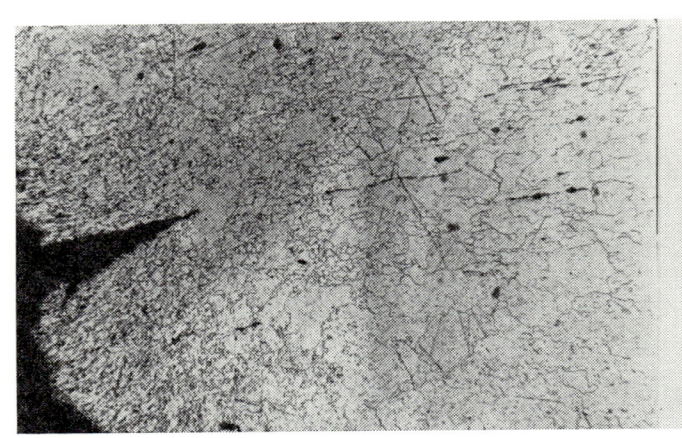

a Longitudinal section showing crack in fracture face, and strings of inclusions, ×80.

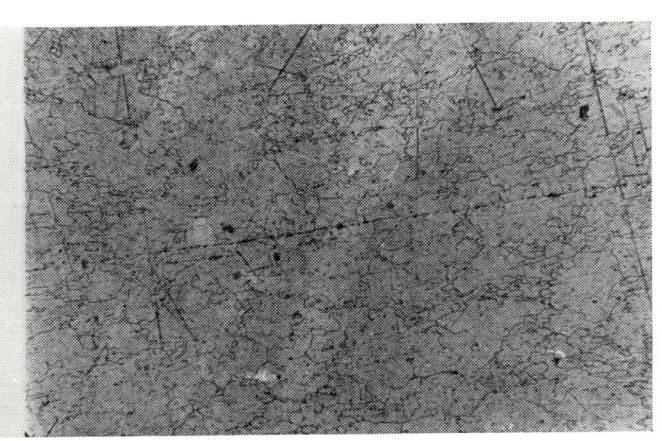

b Longitudinal section showing strings of inclusions, ×80.

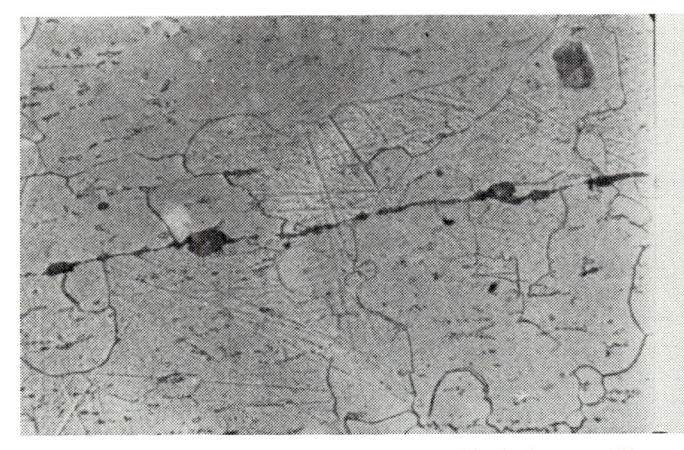

c Longitudinal section showing strings of inclusions, ×450.

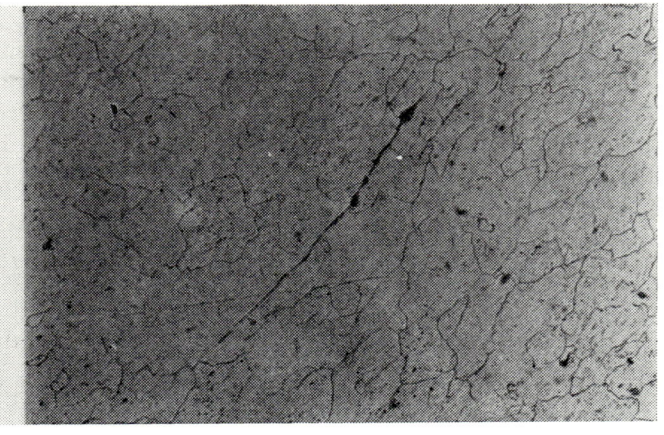

d Transverse section on corner inclusions and cracks, ×160.

Fig. 9.10. Photomicrographs showing distribution of inclusions in cage material

complete break-up, the observed rubbing and wear of the parts of the cage occurred.

The analyses of the grease samples were important, because they suggested that different greases may have been used to lubricate the motor bearings on the site, and as lithium and calcium base greases are not always compatible with each other it is possible that the deterioration in lubricating properties of the grease in this failure may have been accelerated by this means.

CONCLUSIONS

When considering the causes of a premature bearing failure it is vital to inspect all parts of the bearing, its lubricant, and its surroundings as far as possible. In many failures, evidence will be found of several forms of malpractice. Thus, a bearing may show signs of misalignment, electrical pitting, and grease degradation, and yet have failed because of a wrong fit. It is very easy, especially in the rush of industrial life, to see one of these signs and ascribe to it the bearing failure. The authors have found that life is not usually this simple and have learned to expect a complex cause of failure resulting from a very tangled history.

Finally, papers of this sort inevitably draw attention to bearings which fail, but it should be remembered that a grease-packed ball bearing is one of the commonest components in engineering—and one of the most reliable.

ACKNOWLEDGEMENTS

The authors wish to thank the Directors of the G.E.C. and E.E. Co. for permission to publish this paper. They would also like to record their appreciation of the valuable assistance given to them by their colleagues.

Paper 10

SOME PROBLEMS IN THE DEVELOPMENT OF A HIGH-PERFORMANCE GREASE FOR INDUSTRIAL ROLLING BEARINGS

H. D. Moore*   J. W. Pearson†   N. A. Scarlett*

INTRODUCTION

THE ADVENT of lithium hydroxystearate greases is probably the most important development in the field of lubricating greases in the first half of the twentieth century. The early manufacturing and formulation work was intended to produce an economical grease of high performance—chemical and physical laboratory tests being used as yardsticks. It was disconcerting, although not surprising in the light of subsequent knowledge, when field performance in rolling bearings did not reach the high level expected on the basis of the laboratory tests.

At that time a small number of rolling bearing rig tests had just been installed and we had started to compare the new lithium hydroxystearate greases with other commercially available products. The test results confirmed the relatively poor overall bearing performance of both the lithium greases and most of the commercially available greases. It was therefore necessary to revise our previous development philosophy in the light of the field experience and, as a result, certain principles for the future development of greases were determined. These principles may be summarized as follows:

(1) Grease is an engineering component, and as such should be evaluated and developed in the same context as associated engineering components.

(2) This implies that a multi-purpose industrial bearing grease should be evaluated and developed in rolling bearings, not in physical and chemical laboratory tests.

(3) The evaluation in rolling bearings should not be concentrated on one or two types of bearing but should cover all the most important types and the operating conditions found in industrial service.

(4) The grease should be stable in storage, both in the original container and when packed into bearings which are subsequently stored, to ensure that the product eventually operating in the bearings has the designed performance.

We reviewed the bearing types and operating conditions likely to be encountered in service and concluded that it was necessary to cover ball bearings and cylindrical, spherical, and taper roller bearings fitted with a variety of cage designs and running mainly at high speeds and up to a maximum temperature of 130–140°C. Heavy loadings were considered less important as these were likely to be covered by 'heavy duty' industrial greases. We concluded that the evaluation should also include anti-rusting properties of the greases to allow for bearing operation in the presence of moderate water contamination through either condensation or ingress of free water. A special problem arose with the rough running of small bearings, the solution of which was subsequently added to our list of requirements.

The bearing tests already available covered:

(*a*) deep-groove ball bearings running at high speed at ambient temperature and fitted with machined brass cages (Hoffmann Manufacturing Company);

(*b*) a relatively large, high-speed, cylindrical roller bearing running at ambient temperature (Ransome & Marles);

(*c*) a vertically mounted duplex ball bearing running in a relatively overpacked condition (Ransome & Marles);

(*d*) spherical roller bearings running at high speeds and moderate/high temperatures under moderate load (Skefko).

(*e*) deep-groove ball bearings fitted with pressed metal cages running at moderate/high speeds and high temperatures (Institute of Petroleum Method, IP 168; Annapolis).

The MS. of this paper was received at the Institution on 20th November 1969 and accepted for publication on 18th December 1969. 22
* *Shell Research Ltd, Thornton Research Centre, P.O. Box 1, Chester CH1 3SH.*
† *Shell International Petroleum Co., Shell Centre, London.*

Table 10.1. Summary of rig tests used in development work

Rig source	Type of bearing	Test conditions	Requirements for 'pass' result
Hoffmann ('old' high-speed rig)	Two test bearings 1¾-in bore, medium series, single row ball journal. Brass cage located on shoulders of inner ring	Speed: 4000 rev/min Load: 50 lb Duration: 500 h No applied heat	Grease: No excessive change Temp.: Must settle to normal running temperature in under 50 h, preferably under 20 h Bearing: No lacquer deposits or marked cage staining, no wear of bearing parts
Ransome & Marles (MRJ 4E horizontal rig)	One test bearing. Single row cylindrical roller journal. 4-in bore. Brass cage located on shoulders of inner ring	Speed: 2000 rev/min Load: 1000 lb radial Duration: 1000 h No applied heat	Grease: No excessive change Temp.: Peak temperature must not exceed 100°C, temperature must settle to normal running temperature in about 24 h Bearing: No lacquer deposits or marked cage staining, no wear of bearing parts
Ransome & Marles (vertical rig)	One test bearing. Single row ball duplex, 3-in bore. Brass cage located on shoulders of inner race	Speed: 1500 rev/min Load: 600 lb thrust Duration: 1000 h No applied heat	Grease: No excessive change Temp.: Must remain at normal Bearing: No lacquer, deposits or marked cage staining, no wear of bearing parts
Skefko (R2F rig)	Two self-aligning 60-mm bore double row spherical roller journals. Brass cage centred on shoulders of inner ring	*Condition 2* Speed: 2500 rev/min Load: 1872 lb radial Duration: 672 h No applied heat *Condition 3* Speed: 2500 rev/min Load: 1872 lb radial Duration: 480 h Temp.: 45–115°C *Condition 4a* Speed: 1500 rev/min Load: 1872 lb radial Duration: 480 h Temp.: 120°C	Grease: Must show no excessive hardening or softening but some darkening is permissible Temp.: Must remain stable at each temperature level Bearing: Some deposits (but not lacquer) are permissible, no wear allowed (wear can occur in cage bore and on tracks and rollers)
Thornton (heated bearing rig, conditions 1)	One test bearing 1¾-in bore medium series single row ball journal (MTS 14). Brass cage centred on shoulders of inner ring	Speed: 2000 rev/min Load: 30 lb radial Duration: 150 h Temp.: 60–70°C	No lacquer or deposit formation, no chemical attack of bearing parts
Thornton (heated bearing rig, conditions 2)		Speed: 2000 rev/min Load: 30 lb radial Duration: 200 h Temp.: 120°C	No lacquer or deposit formation, no chemical attack of bearing parts
Thornton (multi-bearing rig)	Four test bearings 20-mm bore light series single row ball journal (BRL 020). Pressed metal cage (brass or steel) located on balls	Speed: 5000 rev/min Load: 10 lb axial Duration: To failure Temp.: 135°C	Grease liquefaction, bearing seizure, or point at which temperature becomes uncontrollable is taken as failure point
Thornton (wet bearing rig)	One test bearing. 20-mm ball bearing, plastic cage	Speed: 3000 rev/min Run for 5 h in flowing water. Then stored for three days in covered glass jar half submerged in water	No rusting
Institute of Petroleum (IP 168)	Two test bearings: 40-mm bore bearings (a) Ball, pressed steel cage centred on balls (b) Cylindrical roller, brass cage	Speed: 5000 rev/min Load: 300 lb Duration: To failure Temp.: 135°C	Grease liquefaction, bearing seizure, or point at which temperature becomes uncontrollable is taken as failure point
Thornton (thrust rig)	Two test bearings. 20-mm bore taper roller bearings	Speed: 1425 rev/min Load: 300 lb Duration: 200 h No applied heat	Grease: No excessive change Temp.: Must remain stable Bearing: No marked deposits or wear

The main type of bearing missing from the list was the taper roller bearing. As many of these bearings are used in service a rig was developed for running taper roller bearings under moderate load. As the work progressed, additional rig tests were developed to examine in detail particular aspects of bearing performance. The rig tests are outlined in Table 10.1.

The present paper covers a few of the important problems that were encountered in lubricating the bearings of the rigs, the work done to solve them, and the field trial and service experience subsequently obtained. The purely laboratory tests that were used have been covered elsewhere (1)*, but even in some of those, particularly the storage tests, ball bearings were used as test pieces.

PROBLEMS IN THE DEVELOPMENT WORK

The main problems and their solutions described in this paper are:

(a) Hot running of high-speed bearings.
(b) Deficient anti-rust properties.
(c) Deposit formation at moderate/high temperatures.
(d) Wear of high-speed bearings.
(e) Rough running of small bearings.

Hot running of high-speed bearings

When first operated, a newly packed, high-speed bearing runs hot for an initial period. The temperature later falls to a steady value which is then maintained during subsequent running. Over-packing can result in continuous hot-running but we also encountered prolonged hot-running even with correctly packed bearing assemblies. Prolonged hot-running leads to grease oxidation and/or structural degradation followed by grease liquefaction and leakage or grease hardening and subsequent bearing failure.

References are given in Appendix 10.2.

Two types of hot-running were identified by observing stroboscopically the grease movement in high-speed bearings. In a ball bearing having a bore diameter of $1\frac{3}{4}$ in, fitted with a brass cage centred on the flange of the inner race and revolving at 4000 rev/min, hot-running occurred when the grease originally packed into the bearing failed to 'clear' from the moving parts and attain a settled equilibrium state. The grease continued to recirculate throughout the bearing with consequent continued generation of heat. We were able to relate the 'clearing' characteristics of a grease with its macro-structure; 'smooth' greases failed to clear and ran continuously at an elevated temperature while fibrous or granular greases cleared extremely rapidly, leading to early settled running temperatures. Unfortunately, such rapid clearing tended to result in early cage wear owing to lack of grease in the bearing (Table 10.2). Reference (2) describes the relationship between grease structure observed with an optical microscope and the temperature characteristics in a high-speed bearing. We selected an intermediate structure obtained by close control of the manufacturing process to give an adequate distribution of grease throughout the bearing surfaces in an acceptable period of hot-running.

The second type of hot-running was observed in large roller bearings running at high speeds. Visual observation of grease movement in a 4-in bore cylindrical roller bearing running at 2000 rev/min showed that the grease could clear in an apparently satisfactory manner, but that the work expended in clearing was such that the temperature rose above 120°C with the possibility of dimensional changes occurring in the bearing. Even if the grease cleared satisfactorily and the bearing temperatures appeared to peak at a low value, high running temperatures could still follow because some of the bulk grease in the covers slumped back into the bearing with consequent further working and heat generation. Fig. 10.1 shows

Table 10.2. Effect of grease structure on performance in ball and cylindrical roller bearings

Sample no.	Appearance		Performance in rolling bearings	
	Macroscopic	Microscopic (×200)	$1\frac{3}{4}$-in bore ball bearing, brass cage, 4000 rev/min	4-in bore cylindrical roller bearing, brass cage, 2000 rev/min, peak temperature, °C
7927	Rough, grainy appearance	Highly crystalline	Temperature characteristics satisfactory. Tendency to give cage wear	134
4924	Grainy texture not as rough as 7927	↑ Increasing crystallinity / Decreasing crystallinity ↓	Temperature and wear characteristics satisfactory	127
260	Grainy texture slightly less than 4924		Temperature and wear characteristics satisfactory	77
3331	Smooth texture with slight grain		Temperature characteristics unsatisfactory. Tendency to prolonged hot running	130
7015	Smooth texture, non-grainy	Non-crystalline	Temperature characteristics unsatisfactory. All tests gave prolonged hot-running	—

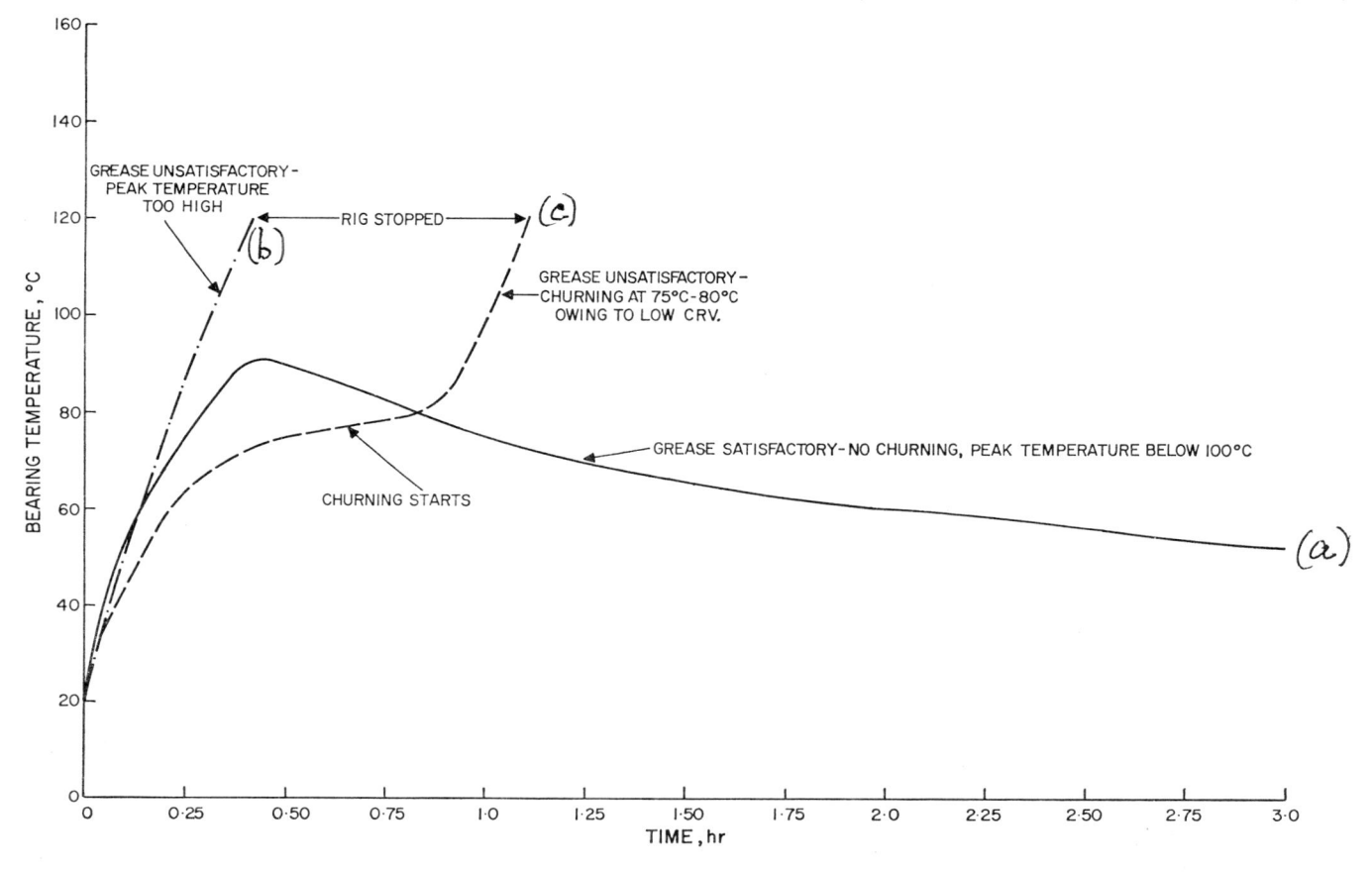

Fig. 10.1. Typical temperature/time curves on 4-in bore roller bearing

typical temperature/time profiles of greases which (*a*) have satisfactory temperature characteristics, (*b*) give a high peak temperature without slumping at lower temperatures, and (*c*) give a high peak temperature because of slumping. We found that the temperature characteristics of greases in the 4-in bore roller bearing were controlled by three factors: viscosity of the base oil, macro-structure of the grease, and yield stress of the grease.

Viscosity of the base oil is the main factor controlling grease viscosity, especially at high rates of shear such as

Table 10.3. Effect of grease base oil viscosity on peak temperature characteristics in 4-in bore cylindrical roller bearings

Grease	Base oil viscosity at 100°F, cS	4-in bore rolling bearing test, 2000 rev/min (Ransome & Marles)	
		Peak temp., °C	Churning below 100°C
499	34·4	84, 85	None
500	81·8	94, 83	None
501	108	98, 96	None
4924	137	127	None
260*	137	77	None

* Modified structure, see Table 10.2.

are found in a rolling bearing. The grease viscosity will determine the amount of work required to clear the grease from the bearing, and hence the peak temperature reached during the early running stages. Table 10.3 shows the effect of varying the viscosity of the oil, in the range 34–137 cS at 100°F with a given grease structure, on the peak temperature reached in the roller bearing rig at 2000 rev/min. For a maximum allowable peak temperature of 100°C (as specified by Ransome & Marles) the maximum base oil viscosity for this particular grease structure was about 112 cS at 100°F.

It was found that peak temperature could also be controlled by modifying the macro-structure. Greases based on an oil with viscosity 137 cS/100°F could then be made to give satisfactory clearability by modifying the manufacturing process to give a rather less granular product. Tables 10.2 and 10.3 include the effect of this relatively small processing change on the peak temperature obtained in the roller bearing rig.

Even with a suitable compromise of base oil viscosity and structure, slumping into the bearing with consequent continued rise in temperature occurred if the yield stress of the grease fell below a certain value during the clearing stages. Reference (3) gives details of the use of Cone Resistance Value (CRV) as a measure of yield stress; the threshold value of CRV was 17 g/cm² for the grease-bearing combination being examined. We found that CRV/temperature characteristics could be modified by changes in processing methods. Thus the manufacturing methods finally adopted had to be selected to give a compromise in grease structure both for satisfactory clearability in ball bearings and large roller bearings running at high speed and also for good CRV/temperature characteristics to minimize slumping in large bearings.

Deficient anti-rust properties

Good anti-rust properties are required to protect the bearings during stationary periods and when running with water present. We thought it necessary to evaluate anti-rust properties under both dynamic and static conditions and developed the 'Thornton wet bearing rig' for this purpose (see Table 10.1). The rig rated marketed greases in the same order as given by service reports.

Uninhibited greases based on lithium hydroxystearate gave poor protection in the rig in the presence of large amounts of water. We examined a wide range of anti-rust additives so that we could select those that ensured good protection without adversely affecting the grease structure which we had so painstakingly developed. Several compounds were promising, but two—sodium nitrite and lead naphthenate—appeared to be exceptionally efficient and economical additives. We finally chose sodium nitrite for reasons that will be discussed later.

Deposit formation at moderate/high temperature

Deposit formation can lead to high starting and running torques, noisy and rough-running bearings, and ultimate failure of the bearings. Deposits form in bearings because of chemical changes in the grease which result in the production of hard or soft oil-insoluble materials. As with all chemical reactions, deposit formation is dependent on temperature and can be catalysed, for example, by the yellow metal of which some cages are made.

We found deposits could be formed in a comparatively short running time at moderate temperatures, e.g. as low as 60°C, and established that this type of deposit was due to decomposition of the glycerol remaining in the grease from the 'in situ' formation of the soap from hydrogenated castor oil and alkali. Although the preparation of the soap from fatty acid, instead of from whole fat, eliminated the glycerol and hence the deposits, the greases were then unsatisfactory lubricants of bearings fitted with brass cages. Oil-soluble anti-oxidant additives proved only

A. Lithium hydroxystearate grease

B. Lithium hydroxystearate grease with sodium nitrite

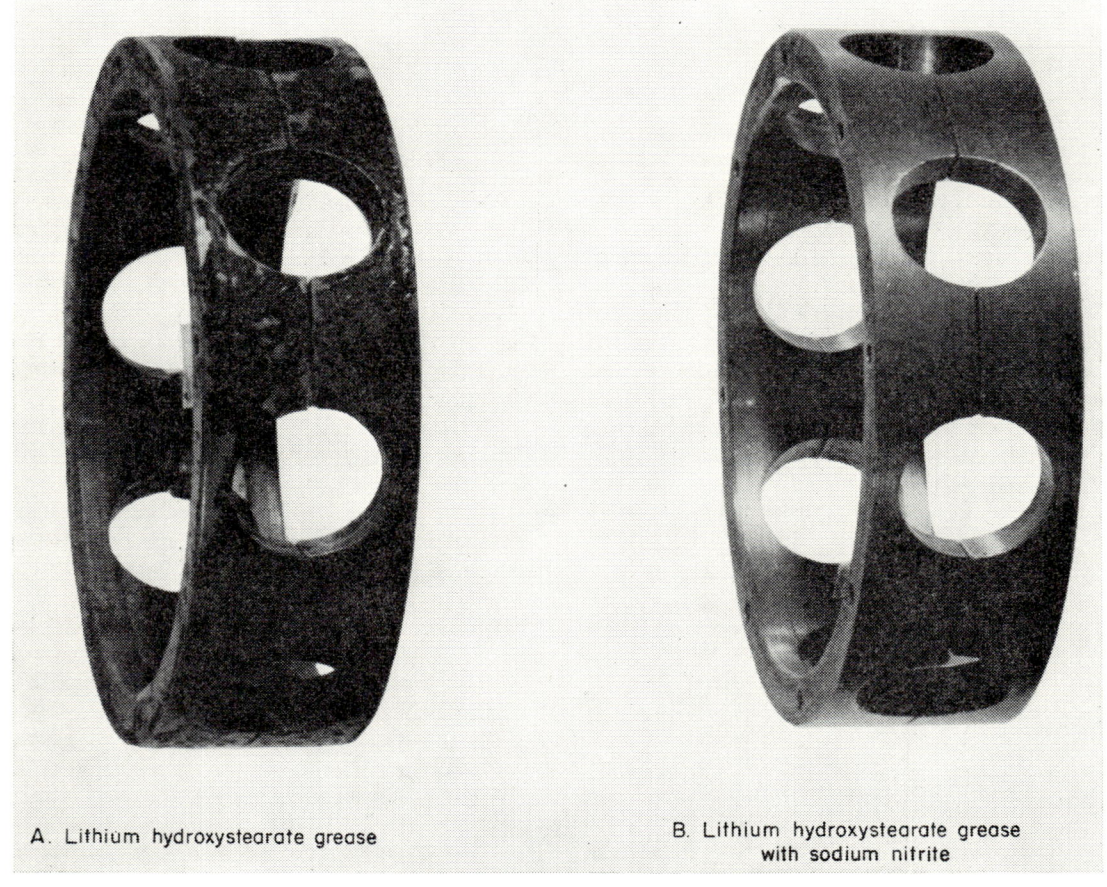

Fig. 10.2. Typical deposit formation on brass cage

pa...lly successful in eliminating the deposits formed ...om glycerol. They were successful with ball and ...drical roller bearings operating at temperatures up to ...t 60°C with correctly packed bearing assemblies, but ...er-packed assemblies, which give considerable grease ... tion with consequent increased deposit formation, ...at higher temperatures, the decomposition of the ...rol still occurred giving deposits and 'coppering' of ...ass cage. Thus, in the Skefko spherical roller bearing ...erating up to 115°C, some hard lacquer-like deposits ...obtained. Furthermore, the deposits at the higher ...ratures (125°C) were increased by oxidation of the ...nd oil, and even greases which did not contain ...ol gave lacquer-like deposits.

... 10.2a shows the type of deposit formed in a bearing ...Thornton heated bearing rig) that was developed to ...e deposit formation. The test had run for 200 h at ... with the bearing in an over-packed condition to ... grease circulation and to accelerate formation of ...ts. Table 10.4 shows the significance of the forma-...f this type of deposit and of cage attack in various ...of bearing run at both moderate temperatures, up ...C, and at high temperatures, 115–135°C. At 135°C ...short life (30 h) was obtained owing to increase of ...acidity in the presence of brass cages followed ...quefaction and grease escape from the bearing ...bly.

...ce oil-soluble anti-oxidants did not prevent this formation of deposits we resorted to the unconventional, oil-insoluble additive, sodium nitrite. This material is soluble in glycerol and completely solved the problems of formation of deposits and attack of brass cages; Fig. 10.2b demonstrates the improvement found. We did not investigate the exact mechanism by which the sodium nitrite acts, but we presume that it forms a reserve of alkalinity which neutralizes the acidic products formed by degradation of the grease (glycerol, oil and soap), thus preventing attack of the brass cage, degradation of the grease, and formation of lacquer-like deposits.

Table 10.4 shows the effect on performance of incorporating sodium nitrite into the grease. There was a marked decrease in general deposits, and a marked increase in grease life at high temperatures, particularly in the presence of brass cages. At 135°C grease life was increased by a factor of more than 20.

For these reasons—i.e. improvement in the performance at high temperatures and minimum reaction with brass cages, together with excellent anti-rusting properties—sodium nitrite was finally selected as our anti-rust additive.

Table 10.4 also shows that the other most promising anti-rust additive, lead naphthenate, which was oil-soluble, did not improve the high-temperature performance or reduce the formation of deposits.

Wear of high-speed bearings

Wear in a bearing takes place where the greatest degree of sliding occurs between surfaces. Whether such surfaces

Table 10.4. Elimination of deposits and improvement of high-temperature performance

Bearing test	Grease A Based on fat and containing glycerol and oil-soluble anti-oxidant A	Grease B Based on fatty acid and containing oil-soluble anti-oxidant A but not glycerol	Grease C Based on fat and containing glycerol and oil-soluble anti-oxidant B	Grease D Grease C plus sodium nitrite	Grease E Grease C plus best oil-soluble anti-rust additive, lead naphthenate	Source of test
(a) 1¼-in bore ball bearing fitted with brass cage, 4000 rev/min/500 h. Runs at about 30°C	Tendency to form gummy lacquer-like deposits	No deposits but tendency to give cage wear	No deposits, slight cage staining	No deposits or cage staining	No deposits, some cage staining	Hoffmann
(b) 1¼-in bore ball bearing, brass cage, 2000 rev/min/150 h at 60–70°C	Gummy deposits, cage staining	No deposits or cage staining	No deposits, slight cage staining	No deposits or cage staining	No deposits, slight cage staining	Thornton
(c) 3-in bore ball bearing, brass cage, 1500 rev/min/1000 h. Runs at about 60°C	—	—	Severe coppering of cage, slight deposits	No cage attack or deposits	—	Ransome & Marles
(d) 60-mm bore spherical roller bearing, brass cage, 2500 rev/min/480 h. Runs at temperatures up to 115°C	—	—	Some lacquer-like deposits and cage staining	No deposits or cage staining	Some cage staining	Skefko
(e) 1¼-in bore ball bearing, brass cage, 2000 rev/min/200 h. Runs at 120°C	—	Severe lacquer-like deposits and cage staining	Severe coppering of cage and lacquering	No deposits, slight cage staining	Severe coppering of cage, lacquering and staining	Thornton
(f) 20-mm bore ball bearing, 5000 rev/min/135°C Life: Brass cage, h Steel cage, h	— —	— —	30 (grease liquefied) 240	810 680–740	250 —	} Thornton
(g) 40-mm bore bearing, 5000 rev/min/135°C Life: Brass cage, h Steel cage, h	— —	— —	120 —	650 >1000	— —	} Inst. Petroleum, IP 168

will be adequately lubricated will then depend upon the relative loads and speeds of the sliding surfaces, the supply of grease to the surfaces, and the lubricating ability of the grease film. With deep-groove ball and cylindrical roller bearings under essentially radial loading, the main sliding surfaces are those between the cage and the shoulders of the inner or outer race (with race-located cages), or between the cage and the rolling elements (with cages centred on the rolling elements). In loaded spherical and taper roller bearings there is also a high degree of sliding between the rollers and tracks and rollers and track lips respectively.

We found that the lubrication of yellow metal cages centred on the flanges of the inner race required the presence of suitable anti-wear additives in the grease, of which glycerol was the cheapest and most efficient. However, when we added sodium nitrite to the grease we found a marked increase in the incidence of cage wear at the same time as we eliminated the attack on the brass cages at both moderate and elevated temperatures. We postulated that the efficiency of glycerol as an anti-wear additive was due to breakdown products, probably acidic, of the glycerol which provided boundary lubrication for the steel–brass combination. Since sodium nitrite is soluble in glycerol and reacts alkaline to acids it would react with acidic decomposition products of the glycerol before they themselves could react with the brass of the cage to provide satisfactory boundary lubrication. Some evidence for this theory was obtained by frictional measurements in a Bowden machine with a steel ball sliding on a brass plate. A solution of sodium nitrite in glycerol gave a substantially higher coefficient of friction than did pure glycerol. Thus, we would expect an improvement in lubrication of the cage bore by an additional anti-wear additive for steel–brass that was (a) sufficiently unstable to break down under boundary running conditions and (b) immiscible with sodium nitrite so that acidic breakdown products would not be immediately neutralized. The additive could not be too reactive towards brass otherwise we would again find unacceptable attack of brass at moderate/high operating temperatures. These requirements pointed to the use of ester type anti-wear additives and not to conventional extreme pressure agents, and one of these additives provided satisfactory lubrication of the cage bore even in the presence of sodium nitrite (4).

In the case of spherical and taper roller bearings, the critical sliding surfaces are steel–steel. We found that the way in which the grease was processed was the main factor affecting grease lubrication of the sliding surfaces in the spherical roller bearings. Fortunately, the process requirements to achieve the desired structure were within the limits we had previously set for satisfactory clearability and CRV/temperature characteristics. Thus we were faced only with tight control of manufacture, and not, as might have been the case, with diametrically opposed processing requirements.

The lubrication of taper roller bearings did not lead to any special formulation or processing needs. Satisfactory performance was obtained within the limits of composition (e.g. oil viscosity) and processing which we wished to use for other reasons.

Rough running of small bearings

Heterogeneous particles such as sodium nitrite are liable to cause rough running, noise, and torque variations in small bearings. We found, as a result of systematic work with particles of varying sizes (Table 10.5), that provided the particle size was below 5 μm, and preferably below 3 μm, acceptably smooth running was achieved.

The sodium nitrite commercially available ranged in particle size up to 500 μm and resulted in very rough running of bearings. Thus we needed to devise methods of incorporating into a grease sodium nitrite with particle sizes below 3 μm.

Various methods for reducing the particle size of sodium nitrite were examined, including the following:

(1) *Air milling.* This gave particles in the range 5–35 μm, which tended to agglomerate.

(2) *Ball milling in a suitable liquid medium.* This was slow and needed a medium of low viscosity and high volatility which posed subsequent processing problems.

(3) *Dissolving the sodium nitrite in a suitable solvent and adding the solution to the hot grease.* This was promising in small laboratory vessels but was much less satisfactory on a large plant scale.

Table 10.5. Effect of particle size of sodium nitrite on grease performance in small bearings

Particle size of sodium nitrite	'Feel' of small bearing when rotated slowly by finger	Running torque at low speed	Running torque at high speed
No sodium nitrite present	Smooth	Low, steady	Low, steady
Up to 200 μm	Very rough	High, variable	Low, steady
Mainly 5–10 μm with some larger particles up to 250 μm.	Rough	—	—
Less than 5 μm but aggregates up to 70 μm .	Fairly rough	—	—
Less than 5 μm but some aggregation . .	Slightly rough*	—	—
In range 5–35 μm	Slightly rough	—	—
Less than 5 μm but aggregates up to 20 μm .	Very slightly rough	Low, steady	Low, steady
Less than 5 μm, mainly 2–3 μm . . .	Smooth	—	—

* Rated borderline by field experience.

We finally devised a more complex but more effective method of obtaining small particles of an oil-insoluble material in an oil suspension (5). A solution of, for example, sodium nitrite in water is made into a 'water-in-oil' emulsion by the use of suitable emulsifying agents and homogenizers. The emulsion is dried rapidly to give a suspension of the solid in oil. This technique gives uniform suspensions of particles in the 2–3 μm range. The suspension in oil is then added to the grease in the appropriate quantity. When the sodium nitrite was added by this technique, the resultant greases gave acceptably smooth running characteristics in small bearings.

FIELD TRIALS AND EXPERIENCE

We had now developed a grease of precise formulation and closely controlled manufacture which adequately lubricated all the types of bearings chosen initially as being important in service, as shown by its passing all the tests listed in Table 10.1—but only in these bearings when run in the laboratory for relatively short periods. This left unanswered the question of whether we had in fact made an advance in terms of performance in the field.

Initially, we tried the grease in applications in which our early 'laboratory high performance' grease had failed. Comparative results in relatively large high-speed bearings confirmed that we had at least achieved part of our objective (Appendix 10.1). Indeed, the new grease gave satisfactory lubrication of these bearings for periods up to 25 000 h, a vast improvement on the 2000 h that we had previously experienced.

Subsequent marketing of the product as an industrial multi-purpose grease in most countries of the world has now confirmed that it gives a very high performance in service. We believe that the field experience has justified the principles on which our development work was based and, in particular, that grease is an engineering component and should be developed as such.

APPENDIX 10.1

PERFORMANCE OF GREASE IN REFINERY ELECTRIC MOTORS

During the early operation of one of our refineries, we experienced considerable trouble with grease lubrication of electric motors in the range 30–150 hp. The main trouble was wear of the cages of the rolling bearings in the motors. The maximum life obtained was only 4000 h before the motors had to be stripped and the bearings replaced.

The motors with which the most trouble was experienced are given in Table 10.6.

Sodium based and sodium/calcium based ball and roller bearing greases gave acceptable peak temperatures and settled quickly but cage wear occurred in less than 4000 h.

A lithium hydroxystearate grease developed on laboratory tests alone, base oil viscosity 137 cS/100°F, ran very hot and the bearing temperature did not settle.

A lithium hydroxystearate grease developed on the principles of this paper, with base oil viscosity 137 cS/100°F, gave acceptable peak temperatures and times to settle.

Long-term trials were initiated on 32 motors including nearly a dozen of the above types. The motor overhaul life has been steadily increased and is now three years, without any change of bearings or relubrication. This corresponds to 20 000–25 000 h and is limited only by the overhaul lives of associated equipment.

Table 10.6

Horsepower	Speed, rev/min	Shaft size, in	Bearings Drive end	Bearings Non-drive end
125	2970	3·5	Ball and roller in one housing	Roller
100	2960	3·0	Ball and roller in one housing	Roller
60	2950	3·5	Ball and roller in one housing	Roller
40	2940	3·5	Ball and roller in one housing	Roller
50	2940	3·5	Ball and roller in one housing	Roller

APPENDIX 10.2

REFERENCES

(1) MOORE, H. D. 'Lubricating grease—laboratory tests and their significance', *Tribology* 1969 **2**, 18.
(2) HUTTON, J. F., MATTHEWS, J. B. and SCARLETT, N. A. 'The optical microscope in the study of lubricating greases and their structure', *J. Inst. Petrol.* 1955 **41** (No. 377), 163.
(3) EVANS, D., HUTTON, J. F. and MATTHEWS, J. B. 'Yield stress as a factor in the performance of greases', *Lubric. Engng* 1957 **13** (No. 6), 341.
(4) MOORE, H. D. and PEARSON, J. W. U.K. Patent No. 773 118.
(5) MOORE, H. D. and RIGBY, R. U.K. Patent No. 778 468.

GREASE LUBRICATION: A REVIEW OF RECENT BRITISH PAPERS

P. L. Langborne*

INTRODUCTION

TAKING A WORLD-WIDE VIEW, the subject of grease lubrication is fairly adequately covered by published literature. The greater part of the published information is of American origin, with Western Europe occupying second place by number of publications. The American scene will be reviewed by another author; this review will deal only with recent British papers.

In view of the small number of papers involved, the term 'recent' has been interpreted in a fairly broad sense and covers papers published over the last five years. In this regard it seemed appropriate to commence with a series of important papers which appeared at the Third Annual Meeting of the Lubrication and Wear Group, held in association with the Iron and Steel Institute at Cardiff in October 1964, which dealt with iron and steelworks lubrication.

For convenience the subject has been considered under the following headings:

Grease properties
Grease tests
Centralized grease systems and bulk handling
Rationalization

GREASE PROPERTIES
Grease properties in bulk

The development of greases for special applications

The environmental conditions obtaining in steelworks are varied and often severe, so that if there were such a thing as the 'ideal' lubricating grease for use in steelworks it would be of considerable general interest.

Harris, Moore and Scarlett (1)† considered the merits and demerits of current products in their pursuit of a grease that could not only lubricate all the various machines found in a steelworks, but could also be dispensed satisfactorily. They listed 17 classes of steelworks plant, giving the corresponding 'critical operating conditions' of temperature, load, and contamination. In only two classes of machinery was 'low temperature' a critical factor; all the others were placed by the authors within the 'high', 'very high', or 'extremely high' temperature categories. Contamination was considered a hazard in six classes.

Harris, Moore and Scarlett then examined methods of grease selection for use in centralized lubrication systems.

After drawing attention to useful earlier work of Evans, Hutton and Matthews (2) on the assessment of slumpability, they discussed problems of pumpability, rightly calling attention to the shortcomings of the S.O.D. viscometer at the low shear rates (below 10 s^{-1}) so often found in centralized systems. Problems of high- and low-temperature lubrication, water resistance, and load-carrying capacity were then discussed.

The authors concluded that lithium hydroxystearate e.p. greases went far towards meeting present requirements but did not 'offer as high a maximum service temperature as is desirable'. Clay-base greases were considered attractive but one difficulty was the impartation of e.p. properties to these greases.

Harris, Moore and Scarlett mention the particular problems associated with lubrication of a Klonne gas holder owing to drying out and hardening of the greases caused by the mechanical admixture of contaminant (elemental sulphur and scale). This problem was made the subject of a paper by Stacey (3), who claimed to have solved the problem by using a non-soap base grease (not further described) which still tended to dry out, but over a longer period. Harris, Moore and Scarlett's solution was to decrease the viscosity of the grease and increase the feed rate, thus washing the contaminants to the bottom of the holder.

An extremely detailed and interesting paper on specialized grease development was that of Armstrong, Balmforth and Berkley (4) which described the evolution of a calcium–lead complex grease, possessing a combination of high drop point and load-carrying capacity together with other improved characteristics.

The MS. of this paper was received at the Institution on 23rd February 1970.
* *Senior Experimental Officer, National Engineering Laboratory, East Kilbride, Glasgow.*
† *References are given in Appendix 1.*

One is apt to wonder what special advantages a calcium–lead complex grease would have over some of the multi-purpose lithium products that have been available since 1947. The authors were quick to point out, however, that it had not proved possible to find a lithium grease with e.p. properties that would withstand high temperatures (separate e.p. and non-e.p. lithium products). It would appear that the key to the success of the calcium–lead grease is the variety of fibre shapes and sizes present in the grease structure as compared to the more uniform fibre dimensions of the lithium hydroxystearate greases. This, said the authors, gave the former a more varied response to mechanical breakdown forces. At high shear rate the calcium–lead grease becomes firmer in consistency (rheopectic) and thus has better sealing properties than lithium greases. An NLGI No. 1 grease of calcium–lead composition is said to have been used in centralized systems.

As a contrast from a grease intended for use in a variety of environments one turns to the search by Jones, Moore and Scarlett (5) for a grease which would effect satisfactory lubrication in one specific environment—within a helium-cooled reactor.

A silica gel–mineral oil grease containing sodium nitrite (5 per cent by weight) and sodium stearate (5 per cent by weight) provided satisfactory lubrication for ball bearings, under simulated reactor conditions, in helium at 200°C. Inclusion of the sodium nitrite was apparently the key factor in this case. Although the sodium nitrite was severely depleted at 200°C the authors suggested that it reacted, either directly or through its decomposition products, with grease components to provide an accumulation of oxidized material in the grease which, in turn, reacted with the metal surfaces to provide an oxide film.

Grease properties in pipe flow

The injection devices used in modern centralized grease systems rely, in many instances, on the grease behaving not merely as a lubricant but also as a hydraulic fluid. In the latter role the grease is caused to actuate valves, pistons, etc., so that grease may be injected and/or the device recycled.

In recent years it has been increasingly realized by both manufacturers and users of centralized systems that grease does not always behave as an ideal hydraulic fluid, particularly in regard to the rapid transmission of pressure. Jackson and Morris (6) draw attention to this phenomenon in their paper concerned with the compressibility of grease in pipelines. Compressibility is found to have an effect on the cycle operating time of certain types of grease systems. The cycle operating time is obtained, in theory, by dividing the total system output by the pump output. However, this time is short in reality because the pump is required to continue in operation *after* delivering a volume equivalent to the system output in order that the system pressure may reach the recycling valve. The magnitude of the extra time so required is determined by the compressibility of the grease. Jackson and Morris, in a carefully controlled experiment, evolved a method of determining grease compressibility and reached a number of conclusions which cannot be fully explored here.

The converse problem, that of achieving a rapid *decay* of pressure in a pipeline, was mentioned by Jost (7) who presented a series of pressure–time diagrams for various types of centralized grease systems. In the single-line self-resetting system it is particularly necessary to know the pressure relief characteristics of the grease in order to calculate maximum pipe length and minimum application time. The German Standard DIN 51805 is valuable in this connection but there is as yet no move to introduce a similar standard in the U.K.

Grease properties in bearings

The performance of grease in bearings has not received the attention that its importance merits, and it is fair to say that until the appearance of the paper by Sims and Hunt (8) there had existed a serious dearth of papers in English dealing with the design of plain journal bearings, lubricated exclusively by grease. Sims and Hunt's paper dealt with:

(*a*) type of grease groove;
(*b*) comparison of industrial greases;
(*c*) friction results and viscous effects;
(*d*) endurance tests;
(*e*) surface finish;
(*f*) friction at commencement of rotation.

If any criticism could be made of an otherwise excellent paper it would be that the authors did not fully describe their experiences with different types of grease groove but, having apparently settled on a suitable groove shape, embarked on a lengthy (though important) process of grease selection. Emphasis was laid on the groove length, which Sims and Hunt thought should extend up to 90 per cent of the bearing width. This contrasts with Continental experience, and with that of the reviewer, which would favour a groove extending over no more than 75 per cent of the bearing width, the mid-third of the groove having a chamfered edge so that the greater part of the grease is induced to leave the groove in the axial centre of the bearing. It has been found that the load thus spreads the grease adequately without excessive end leakage.

On the occasion of the Lubrication and Wear Group's Fourth Convention at Scheveningen (May 1966), Muyderman (9) introduced members to the possibilities offered by the grease-lubricated spiral groove bearing. These bearings can, under certain circumstances, replace porous bearings, ball bearings, or jewelled bearings. Muyderman gave a number of interesting examples of the use of these bearings mainly in connection with small electric motors. The ultimate aim was the factory-filled spiral groove bearing, lubricated for life.

Still dealing with 'plain' as opposed to ball or roller

bearings, Smithyman (10) focused attention on the problem of hot axle boxes in rail wagons—very much in the news of late owing to a series of fires in ammunition trains. According to Smithyman, each wagon would probably incur two hot boxes during its life of 25 years. £600,000 per annum was spent on maintaining plain bearing axle boxes (wagons £420,000 and carriages £180,000). Replacement by roller bearings was proceeding apace but these were by no means trouble-free. Corrosion-fretting was a problem and migration of grease within the box was imperfectly understood.

Scarlett (11) outlined the uses and limitations of grease in rolling bearings and included information on new problems encountered in atomic reactors and aerospace equipment. Cages centred on the inner race are, in general, more difficult to lubricate than those centred on the rolling elements. Bearing design variables were taken into account by plotting speed curves against bearing size, a certain type of bearing being used as a basic reference and factors being applied for other types of rolling element and cages. High speeds were found to reduce life considerably. A 10-fold increase in life was obtained when the maximum speed of a certain bearing was reduced by 25 per cent, and a 50-fold increase was obtained with a 75 per cent reduction at temperatures of about 30°C.

Moore (12) dealt with the grease lubrication of rolling bearings in very basic terms, outlining the importance of correctly charging the bearing and avoiding overpacking.

GREASE TESTS

A concise account of modern laboratory testing methods for grease was that given by Moore (13) as recently as February 1969. Dealing first with three basic tests—penetration, drop point, and water resistance—Moore then dealt with other tests such as oil separation, static oxidation, yield value, viscosity, mechanical stability, volatility, and high-temperature performance. Moore's paper might be criticized on grounds of brevity, but it included a worthwhile list of references together with a table of ASTM and equivalent IP Standards.

A grease test of a very different type was described by Stringer (14) and was of interest because it took no less than three years to perform! The test was made in the course of developing a range of general industrial a.c. induction motors fitted with large-diameter rolling bearings with cases centred on the rolling elements and using a lithium hydroxystearate based grease of No. 3 consistency. The test extended over 22 000 h, the machine being stopped at 1000-h intervals up to 12 000 h and the bearings examined. New grease was added after each examination. After 12 000 h the interval between inspections was increased to 3000 h. Stringer concluded that the useful life of rolling bearings operating, as in this case, at speed factors in the region of 300 000 mm/min is determined by the acceptable limit of the noise emitted by the machine. This in turn is influenced by the machine duty cycle and the rate of regreasing the bearings. A machine allowed to cool periodically will develop 'noisy' bearings at an earlier operating time than a machine which operates continuously.

CENTRALIZED GREASE SYSTEMS AND BULK HANDLING

Centralized grease systems have now assumed such a degree of importance that the provision and selection of suitable grease together with a system design, layout, and operation has become a study in itself. Grease selection has been treated indirectly elsewhere in this review and other aspects of centralized systems will therefore be considered. A number of interesting papers have been written on the subject over the past five years.

Newcomers to centralized grease systems are often surprised to learn that the grease can be expected to perform other functions besides that of a lubricant. Quite frequently the grease is, in effect, used as a hydraulic fluid. On rare occasions the grease is employed as a part of the user's process.

An instance of the latter was given by Ritchie and Robertson (15) in describing the lubrication of the slide tracks of modern sinter machines. A series of pallets carry an ignited charge of coke and ore over a series of windboxes. Fans draw air into the windboxes through the ignited charge. It is essential to keep the cold air leakage through the sliding surfaces on which the pallets run to a minimum. In a previous Lurgi type sinter machine this was done by running the supporting rollers on resilient rails. The sliding surfaces worked at a temperature of approximately 350°C; an asphaltic compound was used as a lubricant and seal.

A recent design uses pallets running on rigid rails but with springs behind the sliding pallet faces. With the redesigned machine the working temperature was reduced to the region of 225°C, enabling a No. 2 consistency lithium grease to be dispensed on to the slide tracks from a centralized system. Grease consumption was about half that of the asphaltic compound.

Jost's paper (16) dealt with the design problems arising during the planning stages of setting up a centralized system including

 (a) poor positioning of lubricating points;
 (b) poor quality standards of piping;
 (c) unsuitable pipe sizing when grease lines were split;
 (d) obstruction by shrouds;
 (e) guards preventing lubrication.

At first sight some of the factors mentioned might appear trivial, but in fact they are very important and failure to attend to them at an early stage could cause a great deal of expense and trouble. The same paper described the arrangements for the bulk handling and distribution of lubricants in a large works, listing 26 grease systems having from 6 to 874 points.

A paper by Stewart (17) dealt with the planned lubrication of a 68-in hot strip mill, which incorporated seven grease systems—five intermittently operated dual line and two continuously operating direct feeding.

Three feed rates were provided:

(1) direct feed to locations where danger of water wash-out existed;
(2) fast feeding intermittent to high-temperature areas;
(3) slow feeding intermittent to the rest of the plant.

It is important to realize that these requirements could not be met in their entirety by any one system. The question is often asked 'Why not employ a super-grease system?'—a system fed with grease in bulk and lubricating an entire works.

Quite apart from the potential dangers of such a system breaking down, the main objection lies in the diversity of feed rates that would be expected of the system. Bulk grease systems can be used to good effect in charging the reservoirs of centralized systems and for certain direct feed applications, but the other feed rates mentioned by Stewart essentially require separate systems.

The design and layout of bulk handling systems was extensively explored in the paper by Hoyland and Brett (18).

A contribution by Jost (7), concerned with pressure-time diagrams for centralized systems, has already been mentioned.

RATIONALIZATION

On the subject of rationalization the paper by Jost (16) was of interest. This dealt with the planning and organization of lubrication for a new integrated iron and steelworks, the Spencer Works at Newport. The lubricants offered by suppliers or specified by plant makers were considered on the basis of functional lubrication requirements. A range of standard lubricants was established and recommendations for special lubricants were not generally accepted if the requirements could be met by one of the preselected standards. As a result of this rationalization it was believed that the number of standard lubricants used was smaller than at any other comparable steelworks in Great Britain. For nearly all (approximately 99 per cent) applications throughout the works only one grade of lubricating grease was used (a lithium grease with lead–naphthenate e.p. agent No. 2 consistency). It would be interesting to know how this installation is faring after an interval of over five years.

A similar concentration on a single lithium base e.p. grease was thought desirable by Stewart (17) describing his experiences in planning the lubrication of a 68-in hot strip mill. The one exception to this practice appears to have been on coiler mandrels where a high-temperature grease was essential.

A single No. 2 lithium–lead grease was also preferred by Hoyland and Brett (18) for use in a bulk handling system.

The contrasting results of Annable and Stafford (19) were of interest. These authors described a scheme for the classification of lubricants for use in a group of integrated steelworks. Their work resulted in a classified list of 14 greases, comprising three cup greases, three ball and roller bearing greases, one high melting point grease, four roll neck greases, one skid grease, one axle grease, and one graphited grease. It was stated that experience in using the classification scheme had, over a period of years, resulted in rationalization, whereby the number of grades had been reduced by over 40 per cent. Although this was presented as a statement applying to lubricants in general rather than to greases in particular, one is left to wonder about the original list of greases before rationalization!

A comparison of Annable and Stafford's paper with that of Jost highlights the difficulties of achieving rationalization in a group of established works as opposed to a new works. Despite these difficulties the rewards of rationalization are considerable, and indeed some degree of rationalization is likely to become mandatory as heavy industry becomes concentrated into larger units.

It is not difficult to visualize a heavy industrial estate of the future in which whole groups of works and factories would be fed with lubricants piped from a central depot maintained by an oil company or group of companies. Lubrication will then be 'laid on' in much the same way as electricity is supplied from a substation. Just as there is no room in electrical practice for supplies at a multiplicity of voltages and frequencies, so there will be no room for a multiplicity of grades of oil or grease. In such a complex, rationalization will be carried to the ultimate.

If such a concept should be thought far-fetched it should be noted that this condition has virtually been reached in the larger steelworks of the present day.

CONCLUSION

In the foregoing review, which does not purport to be in any way comprehensive, the aim has been to mention those developments which appear to have broken new ground. The general tenor of grease papers published over the past five years gives grounds for the hope that, during the next half decade, grease will tend to be approached on fundamental rather than empirical grounds.

Two examples of this trend spring readily to mind: papers on grease testing tend to describe meaningful tests bearing on the functional aspects of the product, whilst papers on centralized systems no longer read like dreary catalogues of manufacturers' fittings but deal with the peculiar behaviour of grease, that mysterious substance so interesting to us all.

ACKNOWLEDGEMENT

The author wishes to thank the Director of the National Engineering Laboratory, Ministry of Technology, for permission to publish this paper. It is Crown copyright and is reproduced by permission of the Controller, H.M. Stationery Office.

APPENDIX 1

REFERENCES

(1) HARRIS, J. H., MOORE, H. D. and SCARLETT, N. A. 'Pursuit of the ideal in lubricating grease for steel works', *Proc. Instn mech. Engrs* 1964–65 **179** (Pt 3D), 42.

(2) EVANS, D., HUTTON, J. F. and MATTHEWS, J. B. 'Yield stress as a factor in the performance of greases', *Lubric. Engng* 1957 **13** (6), 341.

(3) STACEY, F. E. 'Lubrication of a waterless gas holder', *Proc Instn mech. Engrs* 1964–65 **179** (Pt 3D), 168.

(4) ARMSTRONG, E. L., BALMFORTH, N. and BERKLEY, J. B. 'A new calcium–lead lubricating grease for multi-service industrial use', *Lubric. and Wear Fourth Conv., Proc. Instn mech. Engrs* 1965–66 **180** (Pt 3K), 223.

(5) JONES, J. A., MOORE, H. D. and SCARLETT, N. A. 'Grease lubrication of ball bearings in helium', *Proc. Instn mech. Engrs* 1968–69 **183** (Pt 3I), 1.

(6) JACKSON, A. G. and MORRIS, N. R. W. 'Study of the compressibility and transmission of grease in pipelines. Part I. Compressibility of grease in pipelines', *Proc. ISI–I.Mech.E. Conf. Tribology in Iron and Steelworks* 1969 (September), 144 (Instn Mech. Engrs, London).

(7) JOST, H. P. 'An engineering approach to the selection of centralized grease-lubrication systems', *Proc. Instn mech. Engrs* 1964–65 **179** (Pt 3D), 361.

(8) SIMS, R. B. and HUNT, R. T. V. 'Design of grease-lubricated plain journal bearings for a crank-operated cold plate shear', *Proc. Instn mech. Engrs* 1964–65 **179** (Pt 3D), 12.

(9) MUYDERMAN, E. A. 'New possibilities for the solution of bearing problems by means of the spiral groove principle', *Lubric. and Wear Fourth Conv., Proc. Instn mech. Engrs* 1965–66 **180** (Pt 3K), 174.

(10) SMITHYMAN, G. T. 'Tribology in railway engineering', *Proc. Symp. on Tribology in Railways* 1969 (February), 98 (Instn Mech. Engrs, London).

(11) SCARLETT, N. A. 'Use of grease in rolling bearings', *Lubric. and Wear: Fundamentals and Application to Design, Proc. Instn mech. Engrs* 1967–68 **182** (Pt 3A), 585.

(12) MOORE, H. D. 'Lubricating grease. Lubrication of rolling bearings', *Tribology* 1969 **2**, 93.

(13) MOORE, H. D. 'Lubricating grease. Laboratory tests and their significance', *Tribology* 1969 **2**, 18.

(14) STRINGER, L. S. 'A three-year test', *Ind. Lub.* 1967 **19** (6), 234.

(15) RITCHIE, G. O. and ROBERTSON, J. A. 'Special lubrication requirements of iron and steel works', *Proc. Instn mech. Engrs* 1964–65 **179** (Pt 3D), 106.

(16) JOST, H. P. 'The planning and organization of lubrication for a new integrated iron and steel works', *Proc. Instn mech. Engrs* 1964–65 **179** (Pt 3D), 287.

(17) STEWART, J. 'Planning lubrication of a 68-in hot strip mill', *Proc. Instn mech. Engrs* 1964–65 **179** (Pt 3D), 311.

(18) HOYLAND, R. H. and BRETT, R. 'Considerations in the design and layout of grease bulk handling schemes for steel works', *Proc. Instn mech. Engrs* 1964–65 **179** (Pt 3D), 323.

(19) ANNABLE, A. E. and STAFFORD, H. 'The classification of lubricants for a group of integrated steel works', *Proc. Instn mech. Engrs* 1964–65 **179** (Pt 3D), 336.

REVIEW OF RECENT U.S.A. PUBLICATIONS ON LUBRICATING GREASE*

R. S. Barnett†

SCOPE

THIS PAPER SUMMARIZES U.S.A. lubricating grease literature published in the *NLGI Spokesman* (National Lubricating Grease Institute) and the two publications of the American Society of Lubrication Engineers (*Lubrication Engineering* and *ASLE Transactions*). It primarily covers the years 1963 to November 1969, although a few earlier publications have been included because of their outstanding nature. Some preprints from the 1968 NLGI Annual Meeting and the 1969 ASLE Annual Meeting are also covered. Two hundred and twenty-seven publications are reviewed.

Number of papers in each category

Table 1 shows the distribution of effort by listing the number of papers in each main category. It is seen that the classification 'Application and requirements' is ranked first, with 48 papers (21·1 per cent) out of the total 227. The 'Structure and basic research' category ranks second with 31 papers (13·7 per cent), with other categories including lesser numbers.

Table 2 lists the number of papers for each sub-section. For instance, 10 out of 31 papers (32·2 per cent) dealt with 'Theories of structure' in the main classification 'Structure and basic research'.

The author found it difficult to classify certain papers but tried to put them into the category of predominant emphasis. This was particularly the case for some papers which dealt with formulation as well as application.

SUMMARY

Recent lubricating grease research and technology in the U.S.A. has been a wide-ranging endeavour with fundamental research investigations on theories of structure (including complex soaps), permeability, measurements of electrical properties, and electron microscopy adding much to the knowledge of grease structure. NLGI Fellowship studies have contributed significantly to this phase.

Outstanding advances have also been made in manufacture and processing, giving greater efficiencies, shorter times of manufacture, and culminating in the first truly continuous manufacturing process. Radioactive tracing of additive mixing has been accomplished. Packaging has advanced with new materials, and new ingredients for greases have become available. Inventory and quality control has been marked by the use of computers.

Mechanical testing and evaluation has been a major effort in supplementing simple non-bearing bench tests by providing laboratory rig tests using actual bearings and gears. Many new designs of testing equipment have been developed, and studies of the friction, wear and e.p. properties of lubricating greases, including ball joint testing, have been extended.

Analysis and non-mechanical testing endeavours have included a critical look at the significance of the bomb oxidation test, the development of an instrumental method to measure colour, and the use of infrared techniques for quantitative analysis of lithium soap in lubricating grease.

Studies on flow and dispensing have been particularly active resulting in an advanced understanding of grease flow in both theoretical and practical applications. In particular, apparent viscosity determinations have been correlated with flow in pipes in a very useful way. The continuing development of centralized lubricating systems has made for efficiency, economy and safety, and great expansions in bulk handling have come about through better understanding of grease flow.

Formulation of lubricating greases has been marked by development of new thickeners in addition to complex soaps, and in the expanded use of molybdenum disulphide as a solid-film lubricating additive. The development and study of products for extreme environments, including aerospace applications, has resulted in some industrial spin-off. The new NLGI Reference Systems provide samples of the same formulation and manufacture which can be used for research and evaluation by laboratories with assurance of continuity; this being particularly valuable for co-operative investigations by technical

---

The MS. of this summary was received at the Institution on 23rd February 1970.

* *This is a summary only. The full paper may be read in the Library of the Institution at 1 Birdcage Walk, London, S.W.1. The entire bibliography from the paper is printed on pp. 89–93.*

† *Assistant to Technical Services Director (Industrial), Texaco Inc., Beacon, N.Y.*

societies. New discoveries have been made with old ingredients, such as the finding that high-soap-content greases give longer ball-bearing lives. A special grease has been formulated for the food industry.

The application of lubricating greases and requirements for them have been marked by the development of rust-proof applications from the 'inside-out' rather than simply applying an asphaltic undercoating. The new grease-petrolatum-type products applied mainly with airless sprays have given much better rust protection for the life of the automotive vehicle than was obtained previously. Lubrication intervals for automotive chassis greases have continued to increase, although certain makes of cars have reverted to shorter periods and more grease fittings, particularly when conventional greases are used. There has been an intensive development of aerospace applications with special greases such as the perfluorinated types coming into the picture where hard vacuum, high temperatures, and other extreme conditions must be resisted. However, it has been found that more conventional greases may work surprisingly well, not only where sealed against vacuum, but in vacuum environments. Improvements have been made in railroad lubrication, particularly in the understanding of the functioning of journal roller bearing greases and effect of vibration on them, and the trend toward mild e.p. gear-oil traction-motor gear lubricants which give better heat transfer, resist thickening and afford improved lubrication of gears.

The issue of a revised NLGI Glossary containing

Table 1. Statistical comparison of contents

Section	Rank	No. of papers	Percentage
Application and requirements	1	48	21·1
Structure and basic research	2	31	13·7
Mechanical testing and evaluation	3	29	12·8
Formulation and additives	3	29	12·8
Manufacturing, processing, and packaging	4	26	11·5
Flow and dispensing	5	23	10·1
Analysis and non-mechanical testing	6	15	6·6
Production surveys and marketing	6	15	6·6
Classifications, definitions, specifications, regulations, recommended practices	7	8	3·5
Historical	8	3	1·3
TOTALS		227	100·0

Table 2. Statistical comparison within sections

Section	No. of papers	Percentage
I. *Historical*	3	1·3
II. *Structure and basic research*		
Theories of structure	10	32·2
Permeability studies	10	32·2
Miscellaneous	8	25·9
Complex thickeners	3	9·7
Total	31	100·0
III. *Manufacturing, processing, and packaging*		
Manufacturing and processing	11	42·3
Miscellaneous	9	34·7
Packaging	5	19·2
Processing of specific formulae	1	3·8
Total	26	100·0
IV. *Mechanical testing and evaluation*		
Ball and roller bearing testing	10	34·5
Friction, wear, and extreme pressure (e.p.)	9	31·1
Miscellaneous	5	17·2
Noise and vibration	3	10·3
General	2	6·9
Total	29	100·0
V. *Analysis and non-mechanical testing*		
Physical testing	7	46·8
Chemical tests	4	26·6
Miscellaneous	4	26·6
Total	15	100·0

Table 2—contd

Section	No. of papers	Percentage
VI. *Flow and dispensing*		
Rheology studies	8	34·7
Centralized systems	6	26·1
Bulk handling	6	26·1
Flow in pipes	3	13·1
Total	23	100·0
VII. *Formulation and additives*		
Miscellaneous	8	27·5
New thickeners	7	24·2
Extreme environments	7	24·2
Greases containing molybdenum disulphide	6	20·7
NLGI Reference Systems	1	3·4
Total	29	100·0
VIII. *Application and requirements*		
Automotive vehicles	16	33·4
Aerospace	12	25·0
Miscellaneous	8	16·7
Railroads and urban transit	5	10·4
Military equipment	4	8·3
Agricultural machinery	3	6·2
Total	48	100·0
IX. *Classifications, etc.*		
Classifications and definitions	3	37·5
Specifications	3	37·5
Regulations	1	12·5
Recommended practices	1	12·5
Total	8	100·0
X. *Production surveys and marketing*		
Marketing	8	53·4
Production surveys	5	33·3
Miscellaneous	2	13·3
Total	15	100·0

approximately 100 terms now agreed upon in the lubricating grease industry should help users and others to understand these products. The NLGI Grade Classification has been made an American Standard and is being considered for international standardization as part of the joint ASTM/IP Standard for cone penetration.

Production surveys, which are made periodically by NLGI, and recent marketing forecasts have resulted in very helpful statistics and the recognition of trends. Considerable thought has been given to better marketing techniques.

CONCLUSIONS

It is concluded that recent U.S.A. publications in the field of lubricating greases have been marked by great diversity, with major efforts being made in studies of application and requirements, structure and basic research, mechanical testing and evaluation, formulation and additives, manufacturing, processing and packaging, and flow and dispensing, with each of these categories having from 23 to 48 papers and with the total effort here accounting for 186 papers (82·0 per cent) of the 227 publications reviewed.

REFERENCES

All the references from the complete paper are given in Appendix 1.

APPENDIX 1
BIBLIOGRAPHY

(1) ADINOFF, B. 'Grease evaluation for wedge brakes', *Lubric. Engng* 1968 **24** (September), 416.

(2) ALBRIGHT, R. L. 'Bulk distribution of grease', *NLGI Spokesm.* 1963 **27** (May), 59.

(3) ALEKSA, L. A. 'A look at private carriage in industry', *ibid.* 1966 **30** (August), 162.

(4) ALLEN, R. D., DITTER, J. F., GERSTEIN, M. and CHRISTIAN, J. B. 'Submicron size boron nitride as a grease thickener', *Lubric. Engng* 1964 **20**, 339.

(5) ANDERSON, F. W., NELSON, R. C. and FARLEY, F. F. 'Preparation of grease specimens for electron microscopy', *NLGI Spokesm.* 1967 **31** (October), 252.

(6) ANON. 'Bearings conference program', ASLE Annual Meeting, 5th–9th May 1969. (Summary of Dartmouth Conference, 4th–6th September 1968.)

(7) ANON. 'American Standard Lubricating Grease Classification Z11:130–1963', *NLGI Spokesm.* 1964 **28** (May), 41.

(8) ANON. 'NLGI announces a new recommended practice', *ibid.* 1968 **32** (May), 47.

(9) ANON. 'Service industries activity of NLGI', *ibid.* 1965 **29** (August), 137.

(10) ARMSTRONG, E. L. 'Review and appraisal of NLGI Research Fellowship activities', *ibid.* 1964 **28** (September), 168.

(11) ARMSTRONG, E. L. and LINDEMAN, M. A. 'Effects of oil viscosity and soap type on torque in a grease-lubricated journal bearing', *ibid.* 1969 **33** (August), 152.

(12) AZZAM, H. T. 'Friction and wear testing machines to evaluate tomorrow's lubricants', *Lubric. Engng* 1968 **24** (August), 366. (Discussed by J. R. JONES and G. H. KITCHEN.)

(13) BAILEY, C. A. 'Grease on tap—1967', *NLGI Spokesm.* 1968 **31** (March), 422.

(14) BAKER, H. R. and BOLSTER, R. N. 'The effect of thickener purity on the water resistance of a semi-fluid weapons lubricant', *ibid.* 1967 **31** (October), 249.

(15) BARRETO, R. and GONZALEZ, J. 'Characteristics of lubricating greases from calcium complex synthesized in different reaction media', *ibid.* 1966 **30** (September), 190.

(16) BARRY, H. F. and BINKELMAN, J. P. 'Evaluation of molybdenum disulfide in lubricating greases', *ibid.* 1966 **30** (May), 45.

(17) BERG, E. H., HORTH, A. C., NIXON, J., PANZER, J. and PLUMSTEAD, R. J. 'Plugging in centralized systems', *ibid.* 1966 **30** (July), 116.

(18) BERTZ, R. H. 'Modern lubrication methods in the stone industry', Preprint, ASLE Annual Meeting, 5th–9th May 1969.

(19) BERUSCH, A. I. and KING, H. F. 'The Naval Ship Engineering Center's program for improved quiet bearing lubrication', *NLGI Spokesm.* 1967 **31** (September), 196.

(20) BISH, J. M. 'The effect of acidity and basicity of lubricating greases upon their performance', *ibid.* 1968 **32** (September), 193.

(21) BLANK, R. E. and LINDEMAN, M. A. 'Torque tests on a grease-lubricated bearing', *ibid.* 1968 **32** (July), 122.

(22) BLATTENBERGER, J. W. 'Report of the subcommittee on manual test methods and definitions of terms', *ibid.* 1968 **32** (May), 53. (NLGI Glossary, p. 55 of same issue.)

(23) BOEHRINGER, R. H. and TRITES, R. T. 'New aspects in synthetic grease', *ibid.* 1967 **31** (September), 205.

(24) BOLLO, F. G. and WOODS, H. A. In *Advances in petroleum chemistry and refining* (Ed. J. J. MCKETTA, Jun.) 1962, Vol. VI, Chap. 5 (Interscience, New York).

(25) BONER, C. J. *Manufacture and application of lubricating greases* 1954 (Reinhold, New York).

(26) BOOSER, E. R. and GROSSETT, K. W. Discussion of NLGI Annual Meeting Paper on 'Significance of ASTM oxidation stability test for lubricating greases', *NLGI Spokesm.* 1964 **28** (July), 122.

(27) BRIGHT, G. S. 'The use of electrical measurements as an aid in the understanding of fundamental grease structure', *ibid.* 1966 **30** (June), 84.

(28) BRIGHT, G. S. 'Grease transitions—correlation of electrical and viscometric measurements', *ibid.* 1968 **31** (January), 362.

(29) BROWN, W. L. and EWBANK, W. J. 'The effects of thickener concentration on the permeability of lubricating grease', *ibid.* 1965 **29** (June), 77.

(30) BRUNSTRUM, L. C. 'Determinations of grease viscosity from the flow constants', *ibid.* 1963 **27** (August), 144.

(31) BUEHLER, F. A. and RAICH, H. 'Low temperature flow limits for greases', *ibid.* 1967 **31** (May), 49.

(32) BUEHLER, F. A., COX, D. B., BUTCOSK, R. A., ARMSTRONG, E. L. and ZAKIN, J. L. 'Research studies on lubricating grease compositions for extreme environments', *ibid.* 1964 **28** (October), 221.

(33) BUTTLAR, R. O. and CANTLEY, R. E. 'Quantitative infrared spectroscopy studies of lithium soap greases', *ibid.* 1969 **33** (April), 18.

(34) BYRNE, R. 'Performance of greases in railroad journal roller bearings', *ibid.* 1963 **27** (August), 138.

(35) CALHOUN, S. F. 'Current policies and trends in the lubrication of army equipment', *ibid.* 1963 **27** (December), 298.

(36) CALHOUN, S. F. and YOUNG, R. L. 'Rust preventive abilities of grease and their improvement', *Lubric. Engng* 1963 **19** (July), 292.

(37) CALHOUN, S. F. 'Fundamental aspects of grease bleeding', *NLGI Spokesm.* 1966 **29** (January), 328.

(38) CALHOUN, S. F., POLISHUK, A. T. and cooperators 'Frictional characteristics of lubricating greases', Report of ASTM D-2, Tech. Div. G, Sect. IV, Sub-sect. 2, *NLGI Spokesm.* 1969 **33** (August), 164.

(39) CALLAHAN, J. J. and MCDOLE, E. E. 'Comments on NLGI Annual Meeting Paper 'Plugging in centralized grease lubrication systems', *NLGI Spokesm.* 1966 **30** (July), 124.

(40) CARTER, C. F. and BAUMANN, F. 'Gas chromatography of fatty acids applied to grease formulation', *ibid.* 1964 **28** (May), 48.

(41) CAWLEY, P. H. 'A lubricants plant scheduling inventory control, etc. by computer', *ibid.* 1968 **32** (June), 81.

(42) CHAMBERLIN, F. E. and EDWARDS, R. K. 'Two years' experience with a new type of grease manufacturing equipment', *ibid.* 1968 **32** (May), 48.

(43) CHASE, D. L., SANDMANN, L. J. and SAVIDGE, R. E. 'Evaluation of lubricants and coatings for the prevention of thread galling and seizure', Preprint, ASLE Annual Meeting, 5th–9th May 1969.

(44) CHRISTIAN, J. B. and BUNTING, K. R. 'Advanced aerospace greases', *Lubric. Engng* 1967 **23** (February), 52.

(45) CIUTI, B. R., CESARI, M. and BORZA, M. 'Lithium soaps—organophilic bentonite complexes as lubricating grease thickening agents', *NLGI Spokesm.* 1965 **29** (June), 84.

(46) CLARK, M. S. and UNANGST, W. C. 'Bulk grease—a progress report', *ibid.* 1965 **28** (February), 341.

(47) COENEN, C. B. and GORDON, B. E. 'Radiotracer mixing of additives in grease', *ibid.* 1964 **28** (June), 76.

(48) CORSI, A. S. 'General properties of commercial fatty acids', *ibid.* 1963 **26** (March), 378.

(49) CORY, T. C. 'Planning and constructing a new grease plant', *ibid.* 1967 **31** (June), 86.

(50) COX, D. B., OBERRIGHT, E. A. and GREEN, R. J. 'Dynamic and static irradiation of nuclear power plant lubricants', *ASLE Trans.* 1962 **5** (April), 126.

(51) CRIDDLE, D. W. 'Use of an electronic counter to study the size distribution of dispersed grease thickener particles', *NLGI Spokesm.* 1965 **29** (September), 170.

(52) CRIDDLE, D. W. 'Instrumental measurement of the color of greases', *ibid.* 1963 **27** (July), 108.

(53) CRISP, J. N. and ELLIS, W. E. 'Low temperature performance of greases in railway roller bearings', *Lubric. Engng* 1963 **19** (July), 270. (Discussed by R. F. MEEKER and L. LORING.)

(54) CROFT, R. Comments on 'Evaluation of production control methods for predicting storage stability of grease', *NLGI Spokesm.* 1968 **32** (April), 27.

(55) DELAAT, F. G. A., SHELTON, R. V. and KIMZEY, J. H. 'Status of lubricants for manned spacecraft', *Lubric. Engng* 1967 **23** (April), 145.

(56) DEMOREST, K. E. and WHITAKER, A. F. 'Effect of various lubricants and base materials on friction at ultra-high loads', *ASLE Trans.* 1966 **9** (April), 160. (Discussed by D. H. GADDIS.)

(57) DEVINE, M. J., LAMSON, E. R. and STALLINGS, L. 'Molybdenum disulfide diester lubricating greases', *NLGI Spokesm.* 1964 **27** (January), 320.

(58) DICKINSON, H. M. 'Significance of the ASTM dropping point of lubricating grease', *ibid.* 1964 **28** (April), 13.

(59) DITTER, J. F., ALLEN, R. D., THOMAS, H. T., GERSTEIN, M. and CHRISTIAN, J. B. 'Submicron boron nitride as a grease thickener, II. High speed bearing tests', *Lubric. Engng* 1967 **27** (August), 330.

(60) DOOLEY, A. E. 'Lubricants in the food industries', *NLGI Spokesm.* 1967 **31** (April), 18.

(61) DREHER, J. L., KOUNDAKJIAN, T. H. and CARTER, C. F. 'Manufacture and properties of aluminium complex greases', *ibid.* 1965 **29** (July), 107.

(62) DREHER, J. L., CRIDDLE, D. W. and KOUNDAKJIAN, T. H. 'Significance of ASTM oxidation stability test for lubricating greases', *ibid.* 1964 **28** (July), 108.

(63) DREHER, J. L., SMITHSON, W. L. and CARTER, C. F. 'A special grease for the food industry', *ibid.* 1966 **30** (July), 126.

(64) DREHER, J. L. and CARTER, C. F. 'Manufacture and properties of calcium hydroxystearate complex greases', *ibid.* 1968 **32** (November), 293.

(65) DRIVER, J. B. 'Lubrication of airbrake equipment on railroad freight cars', *Lubric. Engng* 1964 **20** (January), 16.

(66) DROMGOLD, L. D., HART, W. and HULME, C. E. 'Automotive vehicle corrosion prevention undercoating', *ibid.* 1964 **28** (May), 42.

(67) ELDRIDGE, H. 'Lubrication requirements of 1964 cars', *ibid.* 1963 **27** (November), 254.

(68) ELDRIDGE, H. 'Lubrication requirements of 1965 cars,' *ibid.* 1964 **28** (November), 261.

(69) ELDRIDGE, H. 'Lubrication requirements of 1966 cars', *ibid.* 1965 **29** (November), 260.

(70) ELDRIDGE, H. 'Lubrication of 1967 cars', *ibid.* 1966 **30** (December), 328.

(71) ELDRIDGE, H. 'Lubrication of 1968 cars', *ibid.* 1967 **31** (December), 332.

(72) ELDRIDGE, H. 'Lubrication of 1969 cars', *ibid.* 1968 **32** (December), 342.

(73) ELLIOTT, H. B. 'NLGI's annual production survey', *ibid.* 1964 **28** (November), 258.

(74) EWBANK, W. J., DYE, J., GARGARO, J., DOKE, K. and BEATTIE, J. 'Permeability coefficients as a measure of the structure of lubricating grease', *ibid.* 1963 **27** (June), 75.

(75) EWBANK, W. J. 'The present status of NLGI work on permeability of lubricating grease', *ibid.* 1966 **30** (August), 167.

(76) EWBANK, W. J. 'The development and present status of the NLGI reference systems', *ibid.* 1967 **31** (April), 14.

(77) EWBANK, W. J. and WARING, R. L. 'Development of a method for determining the leakage characteristics of lubricating grease', *ibid.* 1969 **33** (April), 13.

(78) FARRIS, G. J. and SLATTERY, J. C. 'Flow in an infinite journal bearing', *ibid.* 1963 **27** (November), 263.

(79) FITCH, R. M. 'The how and why of a typical grease plant', *ibid.* 1966 **29** (January), 322.

(80) FITZSIMMONS, V. G., MURPHY, C. M., ROMANS, J. B. and SINGLETERRY, C. R. 'Barrier films increase service lives of prelubricated miniature ball bearings', *Lubric. Engng* 1968 **24** (January), 35.

(81) FLEMING, W. A. 'Space report: past, present, future', *NLGI Spokesm.* 1964 **28** (November), 250.

(82) FORBES, W. W. 'Safe operation of grease plants', *ibid.* 1963 **27** (December), 295.

(83) FRASER, D. J. 'Modern lubricants for paper mills', *Lubric. Engng* 1969 **25** (February), 75. (Discussed by F. J. HANLY.)

(84) FRIEDE, H. M. and SANGSTER, C. T. 'Evaluation of production control methods: storage stability of greases', *NLGI Spokesm.* 1964 **28** (December), 289.

(85) GEBHART, J. C. 'Quality control in compounding and blending lubricating oils and greases', *ibid.* 1968 **32** (April), 15.

(86) GESDORF, E. J. 'Comprehensive review of Grease Dispensing Committee activities', *ibid.* 1965 **28** (February), 352.

(87) GESDORF, E. J. 'Modern methods of lubricant application', Preprint, ASLE Annual Meeting, 5th–9th May, 1969.

(88) GILBERT, D. A. and CATANZARO, F. B. 'Materials, manufacturing and quality control of grease cartridges and their filling and distribution for various markets', *NLGI Spokesm.* 1969 **33** (April), 10.

(89) GILBERT, A. W., VERDURA, T. M. and ROUNDS, F. G. 'Service station grease performance as evaluated in a laboratory ball joint grease test', *ibid.* 1966 **29** (February), 356.

(90) GODFREY, D. 'Friction of greases and grease components during boundary lubrication', *ASLE Trans.* 1964 **7** (June), 24.

(91) GORE, C. E. 'Rustproofing: opportunity for new profits', *NLGI Spokesm.* 1964 **28** (April), 8.

(92) GORE, C. E. 'We have to market if we are going to sell', *ibid.* 1966 **30** (September), 198.

(93) GRAHAM, W. A. 'Efficiencies and costs in U.S. grease manufacture—1965', *ibid.* 1966 **30** (October), 249.

(94) GRAHAM, W. A. 'Grease manufacture around the world', *ibid.* 1965 **28** (January), 313.

(95) GREEN, W. B. and WITTE, A. C., Jun. 'Texaco's continuous grease manufacturing process', *ibid.* 1969 **32** (January), 368.

(96) GUSTAFSSON, O. G. 'The effect of grease lubrication on the vibration of roller bearings', *ibid.* 1967 **31** (November), 289.

(97) HAGSTROM, P. E. 'A forecast of the U.S. grease market 1962–1972', *ibid.* 1964 **27** (February), 350.

(98) HAINES, R. M., MARTINEK, T. W. and KLASS, D. L. 'Theory for inorganic-thickened grease structure—Part III, Apparatus and methods of measurement', *ibid.* 1967 **30** (January), 361.

(99) HARRIS, C. L., READ, J. E. and THOMPSON, J. B. 'Lubricacation in space vacuum—Part 3, Life test evaluation of ball bearings lubricated with oils and greases', *Lubric. Engng* 1968 **24** (April), 182.

(100) HARSACKY, F. J. 'Gear lubrication in today's aircraft', *NLGI Spokesm.* 1963 **26** (January), 318.

(101) HAYER, L. F. 'Arresting fleet corrosion', *ibid.* 1963 **26** (March), 383.

(102) HAYER, L. F. 'The most important preventive maintenance in motor equipment care', *ibid.* 1965 **29** (May), 57.

(103) HERTZLINGER, M. 'Analysis of filled greases using homogeneous solution', *ibid.* 1969 **33** (September), 205.

(104) HIGGINS, W. A. 'Automotive rustproofing compounds', *ibid.* 1966 **29** (March), 381.

(105) HINKLE, C. N. 'Lubricating farm machinery', *ibid.* 1965 **29** (June), 89.

(106) HOGAN, W. T. 'How to read an Industrial Ad', *ibid.* 1968 **32** (July), 128.

(107) HOROWITZ, H. H. and STEIDLER, F. E. 'Calculated performance of grease in journal bearings', *ASLE Trans.* 1963 **6** (July), 239.

(108) HORTH, A. C., SPROULE, L. W. and PATTENDEN, W. C. 'Friction reduction with greases', *NLGI Spokesm.* 1968 **32** (August), 155.

(109) HOTTEN, B. W. In *Advances in petroleum chemistry and refining* (Ed. J. J. MCKETTA, Jun.) 1964, Vol. IX, Chap. 3 (Interscience, New York).

(110) HOTTEN, B. W. 'Distribution of lubricating grease life in ball bearings', *NLGI Spokesm.* 1966 **29** (March), 386.

(111) HOUSE, R. F. 'A modified clay thickener for corrosion resistant greases', *ibid.* 1966 **30** (April), 11.

(112) HOWE, T. B. 'Initial experiments with a ball bearing simulator', *ASLE Trans.* 1963 **6** (April), 133. (Discussed by M. J. FUREY.)

(113) JACKSON, B. E. 'A technique for handling greases and other semi-solid substances', *NLGI Spokesm.* 1965 **29** (September), 179.

(114) JOHNSON, I. H. and HAGSTROM, P. E. 'The 1975 grease market', *ibid.* 1967 **31** (April), 8.

(115) JOHNSON, R. L. and BUCKLEY, D. H. 'Lubricants and mechanical components of lubrication systems for space environment', *Lubric. Engng* 1966 **22** (October), 408.

(116) JOHNSON, R. L. 'A review of the early use of molybdenum disulfide as a lubricant', *NLGI Spokesm.* 1968 **32** (November), 298.

(117) KAUTZ, W. G. 'Antirust: Undercoat—don't undercut', *ibid.* 1963 **26** (January), 324.

(118) KENNEDY, J. V. and GRANQUIST, W. T. 'Flow properties of dispersions of an organo-montmorillonite in organic media', *ibid.* 1965 **29** (August), 138.

(119) KITCHEN, G. H. 'Lubrication of small motor bearings for unattended service in automatic equipment', *Lubric. Engng* 1964 **20** (August), 311. (Discussed by C. E. VEST.)

(120) KITCHEN, G. H. 'Uses of molybdenum disulfide in the communications industry', *ibid.* 1967 **23** (May), 181.

(121) KNOTT, C. R., LINDEMAN, M. A. and POLISHUK, A. T. 'Water spray resistance grease test', *NLGI Spokesm.* 1965 **28** (January), 316.

(122) KOCI, H. H. and BIEN, P. R. 'Design, operation and lubrication of traction motor gears and gear cases', *Lubric. Engng* 1968 **24** (December), 565.

(123) KRUEGER, E. C. and LOCKWOOD, P. B. 'Urban transit lubrication panorama', *NLGI Spokesm.* 1965 **28** (March), 371.

(124) LANE, J. W. 'NLGI's annual production survey', *ibid.* 1967 **31** (August), 164.

(125) LANE, J. W. and FOELL, C. F. 'Let's probe the idea of extended car service', *ibid.* 1963 **27** (April), 23.

(126) LANGMAN, C. A. J., VOLD, M. J. and VOLD, R. D. 'The effect of the liquid component on the penetration of lithium stearate greases', *ibid.* 1967 **31** (August), 152.

(127) LANGNER, F. W. 'New packages for lubricating greases', *ibid.* 1965 **29** (July), 114.

(128) LÅNGSTRÖM, H. O. S. 'A study of the lubricating flow within a grease lubricated anti-friction bearing using "scalped" bearing', Preprint, NLGI Annual Meeting, 29th October–1st November 1967.

(129) LAYNE, R. P. and WARREN, K. H. 'Interaction of lubricating grease with ball bearing vibration in a quiet-running electric motor', *Lubric. Engng* 1966 **22** (August), 302. (Discussed by O. G. GUSTAFSSON and J. E. WEST.)

(130) LEPERA, M. E. 'Petroleum oil characterization using carbon type analysis and infrared spectroscopy', *NLGI Spokesm.* 1969 **32** (February), 400.

(131) LEWIS, P., MURRAY, S. F., PETERSON, M. B. and ESTEN, H. 'Lubricant evaluation for bearing systems operating in spatial environments', *ASLE Trans.* 1963 **6** (January), 67. (Discussions by R. A. BURTON and W. C. YOUNG.)

(132) LIESER, J. E. and WEST, C. H. 'A vibrating rig test for railway bearing greases', *Lubric. Engng* 1968 **24** (September), 399 (discussed by C. R. DANIELS and P. R. MCCARTHY), *NLGI Spokesm.* 1969 **33** (July), 117.

(133) LINDEMAN, M. A., KNOTT, C. R. and POLISHUK, A. T. 'Laboratory grease kettles', *NLGI Spokesm.* 1966 **30** (April), 18.

(134) LINDEMAN, M. A. 'Torque tests on grease lubricated, size 204 ball bearings', *ibid.* 1967 **31** (July), 120.

(135) LIPP, L. C. 'Lubrication of supersonic aircraft', *Lubric. Engng* 1968 **24** (April), 154. (Discussed by F. J. WILLIAMS and C. S. ARMSTRONG.)

(136) LOEFFLER, D. E., CARUSO, G. P. and SMITH, J. D. 'Development and characteristics of microgel greases', *NLGI Spokesm.* 1963 **27** (October), 224.
(137) LYKINS, J. D. 'Lubrication of new 80 inch hot strip mill', *Lubric. Engng* 1966 **22** (September), 350.
(138) MAHONEY, J. E. and KOON, T. J. 'A marketing viewpoint of coal mine lubrication', *NLGI Spokesm.* 1969 **33** (May), 47.
(139) MAHNCKE, H. E. 'What do rolling element bearings need from grease?', *ibid.* 1963 **27** (September), 172.
(140) MARTINEK, T. W. and KLASS, D. L. 'Theory for inorganic-thickened grease structure', *ibid.* 1965 **29** (October), 219.
(141) MARTINEK, T. W., HAINES, R. M. and KLASS, D. L. 'Theory for inorganic-thickened grease structure—Part II, Electrical properties', *ibid.* 1966 **30** (November), 286.
(142) MAYOR, H. A. Jun. and OKON, L. W. 'Is there more to grease classification than consistency?' *ibid.*, 1968 **31** (March), 417.
(143) MCATEE, J. L. Jun. 'Study of dispersants in the preparation of inorganic thickened greases', *ibid.* 1969 **33** (May), 52.
(144) MCATEE, J. L. Jun. and CHEN, L-K. 'Some fundamental aspects of the permeability of organo-montmorillonite greases', *ibid.* 1968 **32** (June), 89.
(145) MCATEE, J. L. Jun. and FREEMAN, J. P. 'Fundamental aspects of the permeability and gel strength of inorganic thickened greases', *ibid.* 1968 **32** (September), 200.
(146) MCCARTHY, P. R. 'Development and evaluation of greases for high temperature, high speed applications', *ASLE Trans.* 1963 **6** (April), 102. (Discussed by J. H. GUSTAFSON and E. L. ARMSTRONG.)
(147) MCCARTHY, P. R. 'Report of ASTM Technical Committee G on dropping point methods for lubricating grease', *NLGI Spokesm.* 1967 **31** (June), 76.
(148) MCCARTHY, P. R. 'High speed, high temperature test rig for grease lubricated ball bearings', *ibid.* 1965 **29** (May), 45.
(149) MCCORMICK, M. M. 'Discussion of the NLGI Meeting Paper 'New aspects in synthetic grease', *ibid.* 1967 **31** (September), 208.
(150) MCKIBBEN, R. F. and FORINASH, D. M. 'Relating humidity cabinet life of lubricants to their service life', *ASLE Trans.* 1963 **6** (July), 233.
(151) MENTON, F. J. 'To buy or not to buy, that is the question', *NLGI Spokesm.* 1969 **32** (January), 374.
(152) MESSINA, J. 'Greases nonreactive with missile fuels and oxidizers', *ibid.* 1963 **27** (September), 177.
(153) MESSINA, J. 'An exploratory study on polytetrafluorethylene thickened greases', Preprint, ASLE Annual Meeting 5th–9th May 1969.
(154) MESSINA, J., PEALE, L. F., GISSER, H. and FRISCH, K. R. 'Lubricants for rapid fire automatic weapons', *NLGI Spokesm.* 1964 **28** (June), 70.
(155) MESSINA, J. 'Perfluorinated lubricants for liquid fueled rocket motor systems', *Lubric. Engng* 1967 **23** (November), 459.
(156) METZGER, P. D. 'Specifying lubricants for a new 80 inch hot strip mill', *ibid.* 1966 **22** (September), 358.
(157) MILES, M. H., MILES, D. W., GABYRSH, A. F. and EYRING, H. 'Stress-relaxation and recovery time for grease and polymer systems. Determination of the relaxation time parameter B', *NLGI Spokesm.* 1964 **28** (September), 172.
(158) MILLER, D. F. 'Chrysler viewpoint—lubrication, 1965', *ibid.* 1964 **28** (December), 294.
(159) MITCHELL, C. H. and SHORTEN, G. A. 'The development of a multi-purpose lubricating grease', *ibid.* 1969 **33** (September), 196.
(160) MOODIE, R. R. 'Preventive maintenance and lubrication in the construction industry', *ibid.* 1965 **29** (April), 11.

(161) MORRISON, W. 'Relationships between composition and rheological properties of lithium hydroxystearate greases', *ibid.* 1963 **27** (August), 145.
(162) NEWMAN, R. H. and LANGSTON, R. P. 'The performance of calcium hydroxystearate greases in wet conditions', *ibid.* 1966 **30** (August), 153.
(163) NIDES, A. W. 'Can our capabilities satisfy your needs?', *ibid.* 1969 **33** (June), 95.
(164) OFFER, L. D. 'Marketing automotive undercoats', *ibid.* 1964 **27** (January), 317.
(165) ORR, J. G. 'What we expect in lubrication from machine tool builders', *Lubric. Engng* 1963 **19** (September), 374.
(166) OSWALT, L. M. 'NLGI's 1964 production survey', *NLGI Spokesm.* 1965 **29** (September), 175.
(167) PANZER, J. 'Application of crystallization theory to the behavior of greases', *ibid.* 1965 **28** (January), 322.
(168) PANZER, J. 'Nature of acetate complexes in greases', *ibid.* 1961 **25** (November), 240.
(169) PAVLOV, V. P. and VINOGRADOV, G. V. 'Generalized characteristics of rheological properties of greases', *Lubric. Engng* 1965 **21** (November), 479.
(170) PLUMMER, E. L. 'Formulation, characterization and performance of aluminium complex imido acid greases', *NLGI Spokesm.* 1964 **28** (August), 142.
(171) POLISHUK, A. T. 'Physical and chemical properties of complex soap greases', *Lubric. Engng* 1963 **19** (February), 76.
(172) POOLER, R. R. 'A new look at traction motor gear lubricants', Preprint, ASLE Annual Meeting, 9th May, 1968.
(173) POPE, C. L. 'Useful grease tests for general plant lubrication', *NLGI Spokesm.* 1964 **28** (July), 104.
(174) RAICH, H., ARMSTRONG, E. L. and PETERSON, B. A. 'Grease manufacture by atomization techniques', *ibid.* 1964 **28** (August), 136.
(175) REBUCK, N. D., STALLINGS, L. and DEVINE, M. J. 'New lubrication vehicle for naval aircraft', *ibid.* 1969 **32** (February), 396.
(176) REDENBAUGH, R. E. 'Development and application of centralized lubrication systems for agricultural machinery', *ibid.* 1969 **33** (August), 160.
(177) REIN, S. W. and MCGAHEY, D. C. 'Predicting grease flow in large pipes', *ibid.* 1965 **29** (April), 20.
(178) REIN, S. W. 'Solution of problems involving grease flow in straight pipe or tubing', *ibid.* 1967 **31** (July), 131.
(179) WOODS, H. A. *et al.* 'Viscosity of lubricating greases at elevated temperatures', *ibid.* 1965 **29** (August), 146. (Report by Section IV, Technical Committee G, ASTM Committee D-2.)
(180) RISDON, T. J. and BINKELMAN, J. B. 'Oxidation stability and anti-friction bearing performance of lubricants containing molybdenum disulfide', *ibid.* 1968 **32** (July), 115.
(181) RISDON, T. J. and SARGENT, D. J. 'Comparison of commercially available greases with and without molybdenum disulfide—Part I, Bench scale tests', *ibid.* 1969 **33** (June), 82.
(182) RITTENHOUSE, J. B., JAFFE, L. D., NAGLER, R. G. and MARTENS, H. E. 'Friction measurements on a low earth satellite', *ASLE Trans.* 1963 **6** (July), 161. (Discussed by P. M. KU and R. L. JOHNSON.)
(183) ROTTER, L. C. and WEGMANN, J. 'The Lincoln ventmeter and its possibilities', *NLGI Spokesm.* 1965 **29** (November), 268.
(184) SAE INFORMATION REPORT J-310a. 'Automotive lubricating greases', *ibid.* 1968 **32** (April), 22.
(185) SARGENT, L. B. Jun. and BUNTING, J. T. 'Grease specifications, now and tomorrow', *ibid.* 1968 **31** (January), 368.

(186) SAYLES, F. S. and cooperators. 'The four-ball EP tester, an ASTM method of test', *ibid.* 1968 **32** (August), 162.

(187) SCHIEFER, H. M., AZZAM, H. T. and MILLER, J. W. 'Industrial fluorosilicone applications predicted by laboratory tests', *Lubric. Engng* 1969 **25** (May), 210. (Discussed by R. S. MONTGOMERY.)

(188) SCHILLING, A. 'Mechanical tests of lubricating greases, Part I', *NLGI Spokesm.* 1967 **30** (February), 388.

(189) SCHILLING, A. 'Mechanical tests of lubricating greases, Part II', *ibid.* 1967 **30** (March), 433.

(190) SCHNEIDERHAN, C. W. 'Fair Packaging and Labeling Act', *ibid.* 1969 **32** (March), 435.

(191) SCHWARTZ, A. A. 'Effect of penetration and thickener content on ball bearing grease performance', *Lubric. Engng* 1962 **18** (May), 237. (Discussed by J. W. JOHNSON.)

(192) SCHWENKER, H. 'Grease lubricants and their potential in aerospace applications', *ibid.* 1964 **20** (July), 260.

(193) SIEGLE, H. J. 'The market for industrial oil and grease in the manufacturing industries', *NLGI Spokesm.* 1968 **31** (February), 389.

(194) SINGH, N. 'Aspect of grease applications and manufacture in India', *ibid.* 1966 **30** (May), 53.

(195) SISKO, A. W. and BRUNSTRUM, L. C. 'Permeability of lubricating greases', *ibid.* 1961 **25** (June), 72.

(196) SLINEY, H. E. and JOHNSON, R. L. 'Preliminary evaluation of greases to 600 F and solid lubricants to 1500 F in ball bearings', *ASLE Trans.* 1968 **11** (October), 330.

(197) SLINGERLAND, I. R. and MILLER, A. J. 'Significance of tests of lubricating greases from the standpoint of an agricultural marketer', *NLGI Spokesm.* 1964 **27** (January), 327.

(198) SPARKS, J. G. and TOVEY, G. R. 'Bulk grease—1962', *ibid.* 1963 **27** (June), 83.

(199) SPICER, F. E. H. 'Performance testing of greases in roller bearing rigs', *ibid.* 1965 **28** (March), 375.

(200) SPRAY, R. P. 'Grease heating systems', *ibid.* 1964 **27** (March), 369.

(201) STALLINGS, L. and cooperators. 'The four-ball wear test—ASTM method No. D-2266', *ibid.* 1968 **31** (February), 396.

(202) STRONG, S. B. 'Grease flow properties and some relationships between them', *ibid.* 1969 **32** (March), 426.

(203) SWANSON, R. F. 'Traction motor grease or gear lubrication in use on the New York Central', Preprint, ASLE Annual Meeting, 9th May 1968.

(204) *Symp. Fluid Lubricants—Their Characteristics, Evaluation and Applications in Hardware*, Abstracts of Papers, C. E. VEST (Chairman), ASLE Annual Meeting 5th–9th May 1969.

(205) TISDALL, R. R. 'Application of corrosion preventive compounds to automobile underbodies', *NLGI Spokesm.* 1963 **26** (January), 318.

(206) TRAISE, T. P. 'Chemistry of polyurea grease thickeners', *ibid.* 1965 **29** (September), 180.

(207) TRITES, R. I. 'Oxidation inhibitor system for extreme temperature greases', *ibid.* 1968 **32** (August), 168.

(208) UHRMACHER, R. R. 'One way to combat the present unsatisfactory profit trend in the farm market', *ibid.* 1963 **27** (July), 112.

(209) VESPER, H. G. 'Have we forgotten how to sell?', *ibid.* 1965 **29** (December), 295.

(210) VITALI, R. and BORZA, M. 'Twist of the fibres of 12 hydroxystearate lithium grease', *ibid.* 1969 **32** (July), 126.

(211) VOLD, M. J. and VOLD, R. D. 'Progress report on the NLGI Fellowship at the University of Southern California', *ibid.* 1965 **29** (May), 54.

(212) VOLD, M. J., UZU, Y. and BILS, R. F. 'New insight into the relationship between phase behavior, colloidal structure and some of the rheological properties of lithium stearate greases', *ibid.* 1969 **32** (January), 362.

(213) WARING, R. L. 'Five years' experience with a new grease plant', *ibid.* 1966 **30** (June), 92.

(214) WATTS, W. A. 'Rigid polyethylene—a new concept in industrial containers', *ibid.* 1968 **32** (June), 86.

(215) WEBSTER, J. C. and EWBANK, W. J. 'The effect of thickener shape on the permeablity of lubricating grease', *ibid.* 1968 **32** (October), 250.

(216) WICKLATZ, E. G. 'New concept of protective coating application', *ibid.* 1963 **27** (October), 222.

(217) WILSON, J. W. Jun. 'Three-dimensional structure of grease thickener particles', *ibid.* 1964 **27** (March), 372.

(218) YOUNG, W. C., CLAUSS, F. J. and DRAKE, S. P. 'Lubrication of ball bearings for space applications', *ASLE Trans.* 1963 **6** (July) 178. (Discussions by D. GODFREY, N. W. FURBY, E. G. JACKSON and R. L. JOHNSON.)

(219) YOUNG, W. C. and CLAUSS, F. J. 'Lubrication for spacecraft applications', *Lubric. Engng* 1966 **22** (June), 219. (Discussed by R. D. BROWN.)

(220) ZAKIN, J. L. and LIN, H. H. 'Permeability of silica greases', *NLGI Spokesm.* 1966 **30** (October), 244.

(221) ZAKIN, J. L., LIN, H. H. and TU, E. H. 'Exploratory studies of the sorption and extraction of additives in lubricating greases', *ibid.* 1967 **31** (May), 43.

(222) ZAKIN, J. L. and TU, E. H. 'Effect of variations in the viscosity and type of mineral oil component on the permeability coefficients of lithium-calcium and baragel greases', *ibid.* 1966 **29** (January), 333.

(223) ZIELINSKI, S. J. 'Lubrication on the farm', *ibid.* 1969 **33** (November), 298.

(224) MURPHY, G. P. Jun. 'Factors that influence grease oxidation and oxidative wear', *ibid.* 1964 **28** (April), 15.

(225) RONAN, J. T., GRAHAM, W. A. and CARTER, C. F. 'New equipment shortens grease processing cycle', *ibid.* 1968 **31** (January), 357.

(226) RUSH, R. E. 'Lubrication and hydraulic practice in a modern steel mill', *Lubric. Engng* 1968 **24** (March), 116.

(227) WEATHERFORD, W. D. Jun., VALTIERRA, M. L. and KU, P. M. 'Experimental study of spline wear and lubrication effects', *ASLE Trans.* 1966 **9** (April), 171. (Discussions by R. B. WATERHOUSE and D. GODFREY.)

Discussion and communications

Mr J. Beardsley (Derby)—When machinery is housed in good environmental conditions, as in workshops, there is little chance of water contamination of roller- or ball-bearing units, but railway equipment, such as rolling stock and locomotives, is likely to operate outside for most of its life and be subjected to extremes of temperature and humidity and to rainfall. Under these conditions, the possibility of ingress of water is increased considerably so that greater attention has to be paid to the corrosion-resisting properties of the lubricants involved.

Experience on British Railways has indicated that unless greases used on outdoor equipment contain sufficient quantities of specially added inhibitors, corrosion of bearings can occur. It has occurred with:

a No. 3 consistency, water-stabilized calcium-base grease, without inhibitors;
a No. 3 consistency sodium-base grease, without inhibitors;
a No. 2 lithium-base grease containing a water-insoluble inhibitor in insufficient quantity.

The first two greases are likely to be equivalent to greases A and C in Table 4.2, from which it may be noted that approximately 15 and 75 per cent 'not suspect' ratings were recorded, or, in other terms, four laboratories would have passed completely grease A and 21 laboratories grease C. By comparison, a British Railways static 'go–no go' test gave failure results on greases of all three types above; furthermore, when the amount of corrosion inhibitor in the third grease was doubled, the product passed the test and no corrosion has since been observed in service. (In the 'go–no go' test grease films, approximately half as thick as the thinnest films found on rollers in service, are prepared on roller-bearing steel discs and subjected to bulk water.)

This evidence means that, on the basis of results from the IP tests, greases which have allowed corrosion of bearings operating under adverse conditions could find their way into service. The test, therefore, is inadequate for some grease users. Among the reasons for this inadequacy are:

no load is applied so that the grease films affording protection are thicker than in a loaded bearing in service;
the test does not take into account the variety of steels used in bearings. These steels, certainly when sodium nitrite is used to prevent corrosion, differ widely in their corrosion resistance.

As it is obviously desirable that a standard test should be of value to all users, does Mr Spicer not consider that the IP method should be re-examined and suitably modified?

Mr R. A. Clarke (Luton)—Following a question put to Mr Spicer which referred to the severity of the test, I wish to confirm that salt water (synthetic sea water) or a 3 per cent sodium chloride solution can be used in the test, but at this stage no definite test data have been obtained in order to state a given repeatability or reproducibility factor. My firm favours the 3 per cent sodium chloride solution because it gives more consistent results. I believe Mr Spicer favours synthetic sea water.

Mr S. C. Dodson (Sunbury-on-Thames, Middx)—In the discussion of yield stress and viscosity of greases in Paper 5 the assumption has been made that, although greases are non-Newtonian, the shear rate between the operating surfaces of a viscometer is constant. This is not necessarily the case. In fact, since a grease has a reduced effective viscosity after shearing and this reduction is greater with higher shear rates, the maximum shear zone will tend to become localized. Eventually the system can be expected to be amenable to mathematical analysis but at present it is necessary to make measurements, as Ir Scholten has, with equipment which is geometrically similar to the equipment in which the grease is used. When we can take results such as those obtained by Yousif and Bogie and apply them directly to the prediction of what will happen in a bearing in service we will really have understood the rheology of lubricating greases.

Mr J. H. Harris (Chichester, Sussex)—There may be some novelty, though not necessarily any value, in offering the comments of one who has been remote from tribological discussions for nearly three years after having been much involved.

The most significant point I should like to make is that some workers with grease still appear to embark on programmes of research on the assumption that they are dealing with perfect, homogeneous compounds. Even under the best laboratory conditions, as I feel sure Dyson and Wilson would admit in relation to the carefully prepared series of greases they used, the dispersion of gelling agent in the liquid phase is not regular or perfect. Fig. 1.4 in their paper illustrates this. It is not improbable that the experiences of Scholten, reported in the opening paragraph of the discussion in Paper 5, may be explained on this basis. Bundles of fibres in soap-based greases, imperfectly broken up in the manufacturing process, may be redistributed in the early stages of shearing, thus 'completing' the making of the grease to a consistence approaching that appropriate to the soap/oil ratio. This stiffening of grease on working is a phenomenon known to grease manufacturers who can deduce from it the extent to which they might obtain more economic dispersion of gelling agent in plant conditions. With prolonged shearing, of course, the characteristic breakage of soap fibres occurs, as shown in Fig. 1.5. Specimens from different batches of the same grade of commercial grease might well show differences of behaviour in the early period of shearing.

It also seems to me that it is still not well understood that greases cannot satisfactorily, for communication of research data, be characterized by reference only to their gelling agent. Even two lithium hydroxystearate greases of the same consistence and with the same oil phase may differ in performance properties, as Moore, Pearson, and Scarlett show in Paper 10. It might be worth while asking the British Standards Institution to devise a code classification of greases for communication purposes by components, roughly on the lines of Fig. 54 in *Lubrication of rolling bearings* (1) with considerable expansion of each class of gelling agent and liquid phase. (The British Standard for calcium-base greases gives an indication as to how petroleum oils may be categorized.) Though such a classification would be far from complete it would at least give a clearer indication than simple references to gelling agent as in Paper 6, where the value of the findings is reduced by absence of adequate data though the authors clearly recognize the significance of oil viscosity and type.

There does seem to be a growing and welcome appreciation of the importance of initial design of rolling-bearing covers to ensure good performance. Paper 3 and three of the report papers show that both established methods and schemes under development considerably reduce risk of failures due to unsatisfactory lubrication, both starvation and excess. Paper 7 by Morgan and Wyllie gives ample evidence of the need for enlightened design and lubrication practice, even where a major advance has been made in the reliability of performance of the specified lubricant. Paper 9 by Yardley and Crump is to be praised for emphasizing the need for an open-minded scientific approach to diagnosis of failures. It has in the past been all too common to blame the grease for failure of a rolling bearing. These authors, as do Morgan and Wyllie, show that many other possibilities or even probabilities require investigation before arriving at a diagnosis.

Finally, I should like to make a comment on 'compatibility of greases' such as the Chairman of Session 2 invited. In my opinion, although it is unwise to change from one grease to another, even though both are considered to be of the same type, without dismantling and cleaning both bearing and housing, the risks arising from such practice are far less than many workers in the field of bearing lubrication suggest. My main contention is that the question should never really arise. If a grease is forced into a housing it may either fill a void in the grease already there or force some of the resident grease into the bearing. If there is already adequate grease in housing and bearing the result may be to force excess into the bearing which may be unable to clear itself sufficiently and thus a rise in bearing temperature may occur. If this happens in a case where a change of grade has been made, the operator may well attribute overheating to 'mixing of the greases', whereas it is evident that it would have occurred if the same grade of grease had been used. This is a typical example of mistaken thinking which leads to fears about 'incompatibility of greases'. As has been demonstrated in fairly comprehensive laboratory work (2)–(4), deliberate mechanical mixing of greases in various proportions leads to changes in significant properties. High-temperature behaviour may be affected and in many cases mechanical stability is lowered to a value less than that of the least stable of the two greases. There probably are circumstances where intimate mechanical mixture of bulk grease may occur, e.g. in a plain bearing or possibly in a taper roller-bearing assembly owing to the pumping action of the taper, but in general one grease fed into a bearing housing containing a different grade will form an interface and no mechanical mixing occurs until the new grease arrives at the bearing itself. It is debatable whether the mixing that may then occur owing to the milling action of the bearing would lead to failure, because even if the mixture becomes fluid this is not objectionable, provided it is retained. As many workers have shown, most satisfactory rolling-bearing greases lose their structure completely in the bearing and form the kind of fluid suspension described by Dyson and Wilson in Paper 1. Nevertheless, my frequently reiterated recommendation in the past has been that change of grade of grease without cleaning and recharging is to be condemned in all cases of bearings running under critical conditions of speed or temperature and inadvisable, for reasons of policy rather than for technical reasons, in other cases.

REFERENCES

(1) HARRIS, J. H. *Lubrication of rolling bearings* 1967 (Shell-Mex and B.P. Ltd, London).
(2) EHRLICH, M. and SAYLES, F. S. 'To mix—or not to mix. A study of wheel bearing grease compatibilities', *NLGI Spokesm*. 1954 **17** (No. 11), 8.
(3) MCLELLAN, A. L. and CALISH, S. R. 'Evaluation of lubricating grease compatibility', *Lubric. Engng* 1955 **11** (No. 6), 412.
(4) MEADE, F. S. 'Compatibility of lubricating greases', Rock Island Arsenal Technical Report 61-2132, 1961. (Distributed by Office of Technical Services, U.S. Dept of Commerce, Washington, D. C.)

Mr D. G. Hjertzen (Luton)—The thrower collars shown on the diagrams in Paper 3 are not usually incorporated in vertical applications and are in fact typical designs of thrower collars used on horizontal machines. When dealing with vertical applications it is necessary to ensure that excessive throwing does not take place and very often the thrower collar is mounted above the bearings. When the thrower collar is mounted below the bearing the outside diameter is often reduced and can have a tapered periphery which mates into a corresponding tapered bore of a retaining cover.

It has been said that it is logical to cover the escape chamber of the escape valve in order to prevent ingress of foreign material, but care must be taken with this technique since it can be very dangerous; it has happened that the cover has not been self-releasing and the grease has built up in the escape chamber so that severe churning has taken place. There are many applications in service where, even though the environment is extremely dirty, it has not been found necessary to introduce a cover.

The illustrations in Paper 3 do not show the use of webs in the housing to prevent the rotation and movement of grease. An interesting development made by SKF is to introduce a segmental web under the grease inlet so that the grease is fed directly into the bearing. This has the effect of reducing the back pressure which can occur and sometimes can result in the seepage of grease through the seals.

Mr J. F. Hutton—I have two points to make on Paper 5: the first concerns the paper itself and the second is a comment on a point raised earlier in the discussion.

Ir Scholten, in the Discussion section of his paper, refers to work done by Bauer, Finkelstein, and Wiberley to lend support to his own observations that the apparent viscosity decreased considerably in the first few minutes of shearing. The experimental results given in this reference ((9) of the paper), however, must be treated with considerable caution because there is some doubt whether or not the viscometer was at all times filled with grease. A Ferranti–Shirley cone-and-plate viscometer was used at rates of shear up to 2×10^4 s^{-1}. Experiments on such an instrument carried out a number of years ago at Thornton Research Centre showed that the torques developed with grade 2 and 3 greases at a moderate rate of shear never reached equilibrium values. Moreover, it was always possible to scrape off grease from the edge of the cone. The conclusion was that grease was being extruded continuously from the shearing gap, leaving behind a partially filled gap. Hence the torque could not be related to any property of the grease. Cox (5) has also reported experiments, using a tracer, which showed that there is a radial component of the velocity of greases in a cone-and-plate viscometer. He also observed extrusion of the greases from the gap. It is necessary for users of cone-and-plate viscometers to check that the gap is always filled with the test material during a measurement.

Extrusion is likely to be associated with elasticity in the grease since it also occurs with highly elastic polymeric liquids, and not with Newtonian liquids. Since elastic liquids can also extrude from concentric cylinder instruments, I should like to ask Ir Scholten if he has any evidence of grease loss from his apparatus, or, a more difficult question, if he is sure that the grease film was at all times continuous.

My second point concerns the existence of a yield stress in a grease. The term 'grease' covers a very wide range of materials from block greases to semi-fluid greases. There should be little doubt that the block greases have yield stresses. As regards the softer ones there is room for doubt. Generally, the yield stress is defined by the test procedure used to determine it, and different procedures give different results. However, a useful simple test for the existence, rather than the value, of a yield stress is to disturb the surface of a volume of grease and to examine it some time later. The time interval may be days or weeks; one assumes that the grease will be sufficiently stable. If the surface levels down to a mirror-like smoothness it can be said confidently that the grease has no yield stress. The stresses on the roughness of the disturbed surface are vanishingly small in the last stages of forming a mirror-flat surface; consequently, there is no doubt about the conclusion. If the surface does not level down then the grease probably has a yield stress, but there is always doubt about it because of the time factor.

Paper 8 is to be welcomed because, together with two preceding ones by Bogie and Harris (6) (7), it is concerned with the application of new techniques to the study of greases. In the past emphasis has been placed on measurements of viscosity and yield stress, whereas in this series of papers equal weight has been given to measurements of viscosity and elasticity.

I should like to ask the authors whether they have made any attempt to fit non-linear models of viscoelasticity to their results?

They make a distinction between the breakdown of structure and the alignment of particles, and they relate features of the rheological results to these processes. It is not clear to me how the relation has been made or how the contributions of the two processes have been separately assessed. There is no indication that the structure and alignment have been studied by direct observation or by any other independent technique. I should be grateful if the authors would amplify their statements on the basic structure of their greases and on the effects of shear on the structure.

I should like to draw attention to links between Paper 10 and two others we have heard this morning. Moore, Pearson and Scarlett pointed out the importance of satisfactory 'clearing' of grease in a bearing and referred to work (8) in which the 'clearability' had been related to the grease structure. In this work it was reported that for a series of sodium-soap greases the types which cleared so readily that they wound out of the bearing immediately on start-up were rough-textured and rubbery. On the other hand the types which did not clear were smooth-textured and buttery. There seemed to be a relation between clearability and rubberiness.

A similar relation appears in the work of Dyson and Wilson presented in Paper 1. They have shown that an elastic polymer solution and certain greases give rise ultimately to thinner films than do Newtonian oils under the same conditions in a disc machine. A sort of clearing process seems to occur more readily with the elastic materials.

The link between these two papers and Paper 8 is the elasticity factor which Yousif and Bogie have set out to measure.

REFERENCES
(5) Cox, D. B. *Nature, Lond.* 1962 **193**, 670.
(6) Harris, J. and Bogie, K. D. *Rheol. acta* 1967 **6**, 3.
(7) Bogie, K. D. and Harris, J. *ibid* 1968 **7**, 255.
(8) Hutton, J. F., Matthews, J. B. and Scarlett, N. A. *J. Inst. Petrol.* 1955 **41**, 163.

Mr R. P. Langston (Cobham, Surrey)—I should like to stress the importance of data from the field. Unless information from laboratory tests and theoretical exercises is able to predict, explain, or at least be compatible with field data it is of little use. I would also stress that the setting-up of a successful field trial involves more than just sending samples of failed components and lubricants to the laboratory. The specialists from the laboratory must visit the site and acquire a background to the use of materials in their natural environment.

With regard to Paper 1, the explanation of why the film thicknesses of grease are less than those of the base oils from which they are compounded suggests that under conditions of oil starvation oils containing viscosity-index improvers could give thinner oil films than could the base oil alone. As the experimental conditions quoted in Paper 1 covered those that might occur in practice for engines and hydraulic systems, this point may have important consequences.

Mr N. R. W. Morris (Hereford)—I wish to comment on Paper 7. The high proportion of bearing failures attributed to dirt and corrosion would seem to indicate that the use of grease to prevent the ingress of contaminants has not been fully appreciated.

There are many industrial applications where, in addition to actual lubrication requirements, additional grease is injected for this purpose. This dictates a more frequent injection period than is normally recommended in which case a centralized lubrication system is often employed to ensure the strict control of injection quantity.

Strict control of injection quantity is most important in respect of rolling bearings because an uncontrolled increase in the quantity injected can result in churning and overheating, although it is significant to note that only one failure is attributed to excess grease against 72 failures due to a deficiency.

Dr S. Y. Poon, B.Eng. (Cambridge)—My contribution refers to Paper 1. Whatever is the mechanism by which a grease is circulated and redistributed in a rolling bearing following a fresh charge of the grease its success as a lubricant must ultimately depend on its ability to maintain an adequate film for the period the bearing is in service. The conventional methods of laboratory test, such as Timken 'O.K. Test', valuable as they are as a means of screening final products under specific conditions, yield little direct information concerning the factors which have led to the eventual failure of the tested bearing. Paper 1, viewed in this light, must be hailed as an important contribution towards our understanding of the grease lubrication in rolling bearings.

At the engineering department in Cambridge we, in collaboration with the authors, have made a separate study of a number of greases, including some of those used by them. An alternative technique was used which measures the change in magnetic flux across the contact zone when a film is introduced. Since the permeability of most non-ferromagnetic material is close to unity, the transducer, once calibrated, provides a direct method of measuring the film thickness of any grease regardless of its composition. The results we obtained confirmed the accuracy of the authors' interpretation of the capacitance transducer.

The fact that a grease gives a thicker film initially is consistent with a viscosity enhanced by the thickener. However, as pointed out by the authors, when it is subjected to continuous shear in the inlet zone of the contact it may lose part of its viscosity owing to mechanical degradation, i.e. the mechanical breakdown of the thickener structure. In order to study the effect of mechanical degradation on the grease-film thickness attempts were made, unsuccessfully, to collect samples of grease from the inlet after various periods of running for micrographical examination. An alternative and perhaps more convincing method was subsequently devised. Samples of two lithium hydroxystearate greases, designated GL2 and GL3, were put in a Klein mill and subjected to shear for a prolonged period (one week at 60°C) until a substantial proportion of the thickener had been broken down from the typical long twisted fibres shown in Fig. 1.4 to those shown in Fig. D1. The results of the film thickness of the ex-Klein mill samples are shown in Fig. D2a and D2b. The effect of the thickener breakdown on the film thickness is immediately obvious when comparison is made with the results of the virgin samples tested under similar conditions.

When the experimental points of any set of the result are moved bodily towards the right by a suitable amount (equivalent to a shift of about 7 min on the time scale) they fall closely on to the curve of the virgin sample. It confirms the view that in the *initial period* the time-dependent behaviour of the film thickness is mainly controlled by the mechanical degradation. This is similar to the mechanical degradation of polymer-enriched lubricants reported in the literature, e.g. (**9**). It is of interest to note that it took only 7 min to accomplish in an elasto-hydrodynamic-lubrication (e.h.l.) contact what a Klein mill would take a week to do; in other words the severity in the inlet zone of an e.h.l. contact exceeded the severity of a Klein mill by a factor of 200.

The authors suggested that partial depletion of the grease from the inlet zone might account for a grease film's being thinner than its base oil in a continuous test. Direct observation of the inlet meniscus by an optical method was made and the results for the grease and its base oil

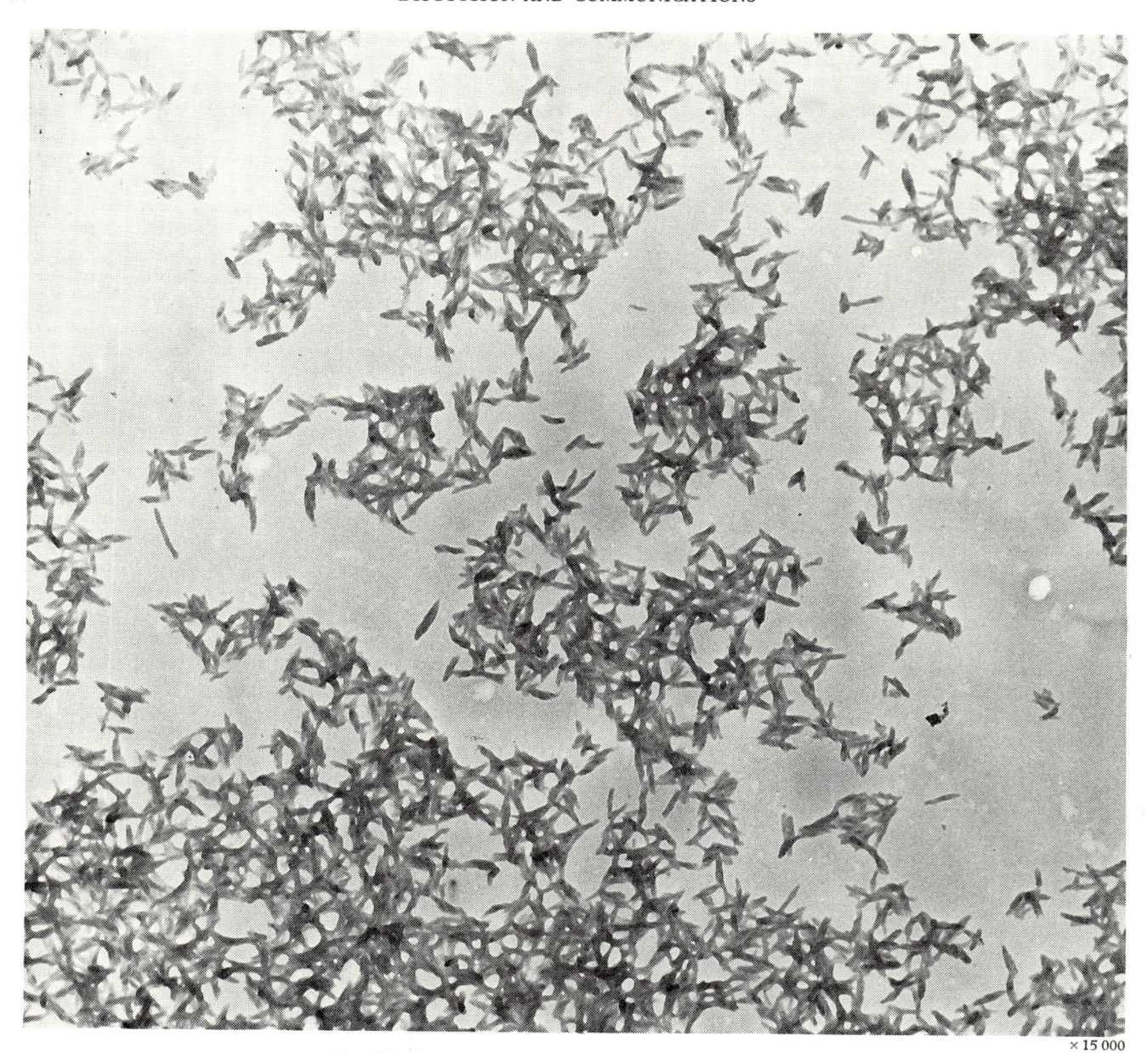

Fig. D1. Micrograph of GL4 (ex-Klein mill sample)　×15 000

are shown in Fig. D3 as a function of time. It would seem that, since the meniscus of the grease remains above that of the base oil, it is difficult to explain simply by depletion the variation of the grease-film thickness when it has fallen below the corresponding oil-film thickness. However, the fact that the *variation* of the inlet meniscus is of the *same time scale* as that of the grease-film thickness suggests that there may exist a connection between the two. The following observations provide some further evidence.

In the course of a test it was observed that the grease showed a tendency to collect gradually at the outlet of the contact zone and form into streams separated by meniscus. If the disc machine was then stopped the grease at the rear of the contact was carried around the circumference of the discs and placed in the front. When the machine was restarted the film thickness registered some gain compared with the thickness immediately before the stop. No such increase of film thickness was observed if the excessive grease at the rear was carefully removed before the stop-restart procedure, which was the same as the No. 3 operating procedure described in Paper 1. Thus results shown in Figs 1.14 and 1.15 obtained by this procedure need to be interpreted in the light of the above observation.

The study by the authors has undoubtedly modified our view of grease in e.h.l. On the other hand the problem of grease in e.h.l. cannot be fully resolved unless such ques-

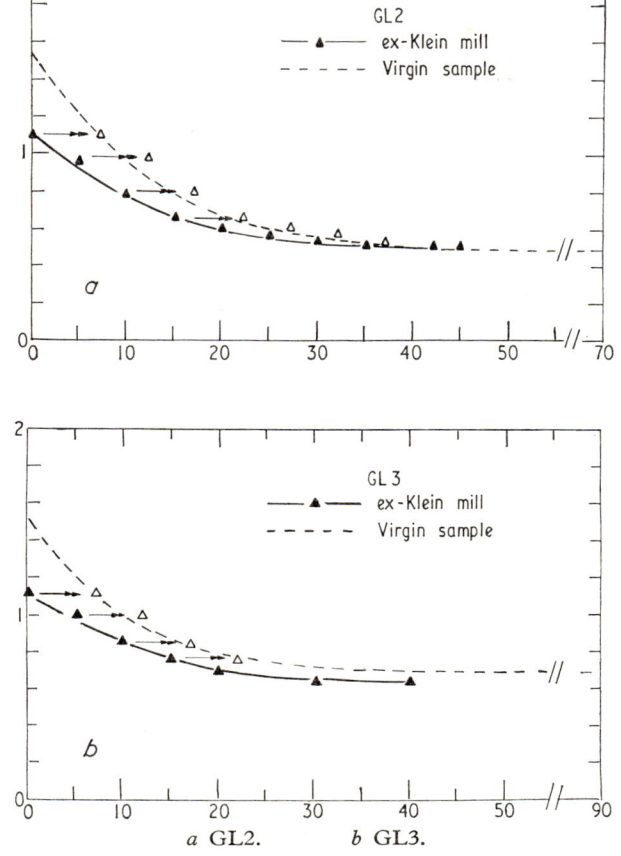

Fig. D2. Variation of film thickness with time

tions as the dynamic equilibrium of the free boundaries and the effect of the boundary conditions on the film thickness of a suspension in e.h.l. are appreciated.

REFERENCE

(9) HAMILTON, G. M. and ROBERTSON, W. G. 'Lubrication of rollers with oils containing polymers', *Proc. Instn mech. Engrs* 1966–67 **181** (Pt 3O), 192.

Mr G. Shorten (Hanley, Stoke-on-Trent)—Mr Sallis and Mr Wilson (Paper 6) state that p.t.f.e.-filled greases have useful low-friction properties but improved resistance to wear is unexceptional. My firm is interested in p.t.f.e.-filled lubricants and find that p.t.f.e. containing carbon fibre and/or conventional solid lubricants such as molybdenum disulphide, graphite, and tungsten disulphide do improve load carrying and resistance to wear. Have they any information on filled p.t.f.e. lined or impregnated bearings?

I cannot agree with their comments on non-availability of polyethylene-treated greases in the U.K. This class of lubricant to the specifications mentioned is made, and has been supplied for a number of years, in this country.

On p. 51 of Paper 7 it is stated that 'XG-274 already meets one of the most severe rust-prevention tests in use and it would not be easy to obtain or specify a higher standard', and on p. 56, 'there is no real prospect of... increasing the rust-preventing properties of the grease'.

As the test used is static and uses distilled water, would not the IP 220/67 test using synthetic sea water be of interest, and has the question of improving XG-274 in this respect been raised with British suppliers of lubricating grease?

On p. 55 it is stated that 'the percentage of bearings lubricated with XG-274 which contained dirt was higher than the percentage of bearings containing dirt and lubricated with other greases'. Would not this indicate that the sealing properties of this XG-274 are inadequate in comparison with other greases, and could this be due to the use of low-viscosity oils to meet the low-temperature torque requirement resulting in a deficiency in grease structure, i.e. long-fibre formation?

Where the dirt has been introduced with the grease during relubrication, has any consideration been given to the use of grease guns using cartridges, thereby dispensing entirely with tins or tubes of all sizes? This method ensures that the grease, as supplied, is contaminant-free when fed into the bearings.

Having followed a similar development programme (10) to that described in Paper 10, my firm has found that the formulation best suited with satisfactory yield stress, macrostructure, and apparent viscosity yielded a 60-stroke penetration of 245–265. This does not fit the NLGI system and we found resistance from both salesmen and customers to accept this range. Have the authors any comment on this aspect, or have they found any correlation on yield stress by alteration of apparent viscosity of the greases on rig testing?

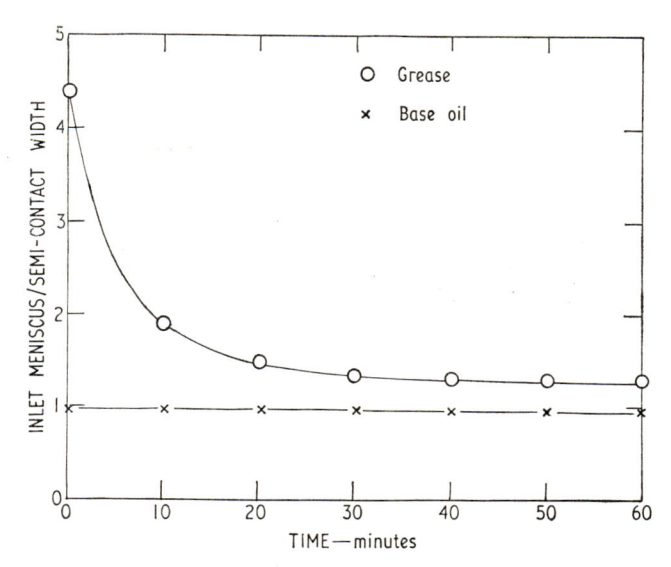

Fig. D3. Variations of the length of meniscus in the inlet zone with time for a typical grease and its base oil

REFERENCE

(10) MITCHELL, C. H. and SHORTEN, G. A. 'The development of a multipurpose lubricating grease', *NLGI Spokesm.* 1969 **33**, 196.

Dr D. Summers-Smith, B.Sc., C.Eng., M.I. Mech.E.—I wish to comment on Paper 3. The acceptance of 'Standard Life' (ISO R/281) as a basis for design with rolling bearings with respect to load and speed has directed attention towards the lubrication arrangements; for clearly lubrication must not become the limiting factor on life. There have been very welcome recent publications by two rolling bearing manufacturers (11) (12) and one grease manufacturer (1) (13) giving methods of estimating grease life in rolling bearings or, more relevant in the context of Mr Mullett's paper, grease re-lubrication intervals.

The basis for grease life prediction is purely empirical and it is of interest to the user, who is not tied to one bearing manufacturer nor to one grease supplier, to see how far the different methods agree. For a 3 in (75 mm) medium-series deep-groove ball bearing, the estimated re-lubrication intervals for 3000 rev/min, at a temperature of 70°C, are as follows:

Method	Re-lubrication interval, h
Hoffman	2600 (standard life 40 000)
Hoffman	2100 (standard life 25 000)
SKF	2000
Shell	2700

These values agree reasonably well with the figure of 1500–2000 h given in Paper 3. The significance of a re-lubrication interval of this order is that, if the grease life is assumed to be twice the re-lubrication interval, clearly 'packing and leaving' is not acceptable on a plant with an overhaul period of a year or more (as is typical of the process industries) and a 'grease valve' or some alternative method of lubrication is essential.

The data on grease valves for paired rolling bearings given by Mr Mullett are thus very valuable. There are, however, two points that are worth noting. The flinger disc acts as a very efficient pump for grease and can, in a vertical application with the disc mounted below the bearing, completely clear the bearing of lubricant, particularly if there is some vibration that causes the grease to slump. This feature would tell against a number of Mr Mullett's recommended arrangements. Perhaps the reason they were satisfactory in the tests is that grease additions were made at intervals of 12 h. It would be interesting to know if Mr Mullett has satisfactory experience with these designs in practical applications when re-lubrication was only carried out at infrequent intervals.

The second point is closely related; on p. 3 Mr Mullett recommends a grease injection at intervals of 300 h.

In the process industries with continuously running plant the need for such frequent attention is not acceptable and where a grease re-lubrication interval of less than 500 h would be necessary an oil-lubrication system is preferred. If a very large safety factor is being built into these recommendations, this should be clearly stated, otherwise it is bound to tell against grease lubrication.

Grease life or re-lubrication interval is now becoming a critical feature in the design of rolling-bearing arrangements. Some standardization of the method by which it can be calculated would thus be of great benefit to machine designers and machine users.

REFERENCES

(11) HOFFMAN MANUFACTURING COMPANY LTD. 'Grease-relubrication intervals', *Ind. lubric. Tribology* 1969 **21**, 304.
(12) SKEFCO BALL BEARING COMPANY LTD. *Engng Bull. QEU 690120* 1969.
(13) MOORE, H. D. 'Lubricating grease: lubrication of rolling bearings', *Tribology* 1969 **2**, 93.

Mr J. K. Vose, B.Sc., C.Eng.—I was particularly interested in Mr Mullett's Paper 3 because we were faced with a similar problem in the development of the larger end of our standard metric-motor range. Our solution most resembles Set-up 5 (Fig. 3.5) and, as most electric motors have horizontal shafts, is not usually subject to the stated speed restriction. The specific comments I should like to offer are as follows:

I remember some years ago reading a patent taken out by a German electrical manufacturer (AEG or Siemens, if my memory serves) claiming protection for one or more forms of two-bearing grease-valve type pressure relief.

It is sometimes suggested that foreign matter can too easily gain access to the bearings in dirty environments. This adverse criticism probably is a result of cases in which there has never been sufficient grease packed into the housing, but in any event it is fairly easy in general to arrange some protection for the grease ejection port.

In his presentation Mr Mullett referred to the undesirability of bends in the grease ejection passage. My own experience has been that there is no real objection to bends provided that they occur near to the rotating grease valve and provided that there is a significant flare in the section of the passage.

Although Mr Mullett refers to his preference for the addition of grease whilst the machine is running, I would tend to put the ability to add grease whilst stationary or running as a very worth-while advantage which might be listed with the six advantages which Mr Mullett has included in his Introduction.

Pressure relief using grease valves was introduced on a commercial scale for AEI electric motors in the 1950s. In the early days we recommended lubrication procedures which involved watching for the egress of grease. Although suitable where only a few machines are concerned, procedures of this kind tend to be wasteful both of grease and, more importantly, of time and we have modified our instructions to call for injections of fixed quantities of grease at fixed intervals. This also covers the case of an installation with both running and stationary machines where, for simplicity, all the bearings are lubricated at the same time.

I think some mention might be made of the desirability in some cases of positioning grease inlet and ejection points to suit the prevailing temperatures. Particularly in outdoor machines which are lubricated when not running it is desirable to have the grease valve in an area which becomes warm quickly because otherwise there is a risk of grease leakage through, for example, shaft running seals before the grease valve becomes operative. This does, perhaps, apply more to single-bearing arrangements but is worthy of mention here.

I turn now to Paper 7. From the standpoint of an electric-motor manufacturer the bearing trouble experienced by MOD(N) is greater than would be expected with industrial motors (it must be remembered that a *bearing* failure rate of about 4 per cent represents a significantly higher *machine* failure rate) even though, because of shock-loading considerations, the bearings tend to be of larger size for a given output. I believe that there are several reasons for this state of affairs and that an improvement could be obtained from the adoption of rotating grease-valve pressure relief. At the present time, motors for MOD(N) have space left in the bearing-housing caps for the grease expelled from the bearings in the process of clearing. Because of the relatively long periods of testing undergone by these motors there will be only a thin film of grease on the working surfaces of the bearings when the motors are dispatched from the manufacturer's works. During the long time between the testing of the motor and its commissioning, i.e. the period during which the motor is subjected to the adverse atmospheric conditions of the shipyard and the vibration due to shipyard activities, the protection to the bearings offered by the grease must be minimal. Furthermore, when the time comes for the starting of the motor it is likely that the lubricating properties of the grease will have seriously deteriorated with a consequent risk of failure during a critical period. If pressure relief were adopted the bearings and housings could be completely charged after testing of the motor with the knowledge that this would not cause harm to either the grease or the bearings. Long-term reliability would result from the facility offered for 'in service' addition of lubricant during routine maintenance.

The number of failures due to excessive axial load and turning of the outer race in the housing may reflect the difficulty of manufacturing a two-ball-bearing motor, where one bearing outer race must slide axially without undue restraint whilst not being so slack in the housing that rotation ensues. It has been suggested that a freer fit could be safely used if an elastomeric O-ring were put in a groove in the bearing seating of the housing, but I have no knowledge that this scheme has ever been used commercially.

The washing out of a new 'as received' bearing has often revealed a fair crop of debris which is, of course, in the best position to cause bearing-track damage. It is to be expected that improved cleaning techniques will produce some marked improvement.

Finally, I wish to comment on Paper 10. In my own work on the assessment of greases for use in relatively large high-speed rolling bearings I have often experienced difficulty in obtaining adequate field service data and I believe that this problem has been encountered by other investigators in this area, including those of oil companies other than Shell. Although bearing arrangements similar to those referred to in Table 10.6 were used with apparent success, or at least acceptability, in pre-1939 electric motors it was quickly found that failures were all too common when motors of generally similar construction were supplied in the post-1945 period to industries, such as petroleum refining, demanding continuous operation for long periods at, or near to, maximum output. We quickly decided that oil lubrication must be adopted as the only sure way, at that time, to avoid further trouble. Shell had been—unfortunately as it seemed at the time—the recipient of a good proportion of the grease-lubricated machines concerned. I now find some consolation in the thought that without those motors Shell might have found difficulty in gaining such useful, and convincing, field experience on difficult-to-lubricate bearings! Refinery maintenance engineers, understandably, often appear loath to allow 'boffins' the facilities to 'experiment' with machines which are already operating satisfactorily.

Authors' Replies

Mr G. W. Mullett—I appreciate Mr Harris's remarks concerning the importance of the correct design of rolling-bearing covers to ensure good performance. Often one sees covers of peculiar shapes, with capacities either too large or too small, with over-elaborate sealing arrangements and with inefficient provisions for regreasing. In some cases, mostly in machine interiors, the necessity for covers has been ignored. This leads me to think that a useful contribution to the art would be a paper, to the Institution, developing the principles behind cover design and setting out some practical rules for guidance.

I have read Mr Hjertzen's remarks with interest. My paper does illustrate the relative merits of thrower collars (valve discs) either above or below the bearings; the former restricts the permissible speeds in vertical mountings. The point made that the disc above is sometimes preferred, because excessive throwing is caused when it is below, is, I think, invalid. I prefer a below mounting. I did remark in my paper that when it is below the disc can be of a smaller diameter.

I have tested discs with tapered peripheries and found them to be quite successful below a bearing but not above. Indeed a properly proportioned flat disc with cylindrical edges placed beneath a bearing is also successful. If one goes a stage further, in a mounting such as Fig. 3.6, where the end cover is so shaped that it forms a convenient receptacle, the set-up works successfully without a disc at all, the purged grease falling into the cover, which must, of course, be regularly cleaned out.

I wholly agree that covering the escape chamber (eject port) can be very dangerous. The case I quoted, kindly described to me by an engineer with the National Coal Board, related to a bearing failure caused by the ingress of coal dust. The machine vibrated severely and I judge that the escaped grease blob was shaken from its position, leaving the port open for the dust to enter the mounting. I think the vibration was the prime cause of the failure. However, dry powders, such as coal dust, if they coat unprotected grease pads, do have a drying effect and the pads become of the consistence of dough. In this state, their adhesion is weakened and a pad can fall, as a lump, from the port end, particularly if vibration is present.

I am aware of the use of segmental webs but I have had little experience of them and cannot really comment. However, we experienced no injection troubles due to back pressure. I think this is because of the wise choice of our inject position—into a corner of a recess which was filled with grease on assembly and which would remain full, see Figs 3.1–3.8.

I welcome Dr Summers-Smith's comments. The virtual abandonment of the old and convenient principle of 'packing and leaving' until the yearly overhaul comes round is the price one pays for the up-rating of modern machinery. Speeds and temperatures can only be increased to meet the requirements of our time by the introduction of re-lubrication procedures. I agree that lubrication life today is more important than fatigue life and it has been for some time past. From this aspect, Dr Summers-Smith's time of a 500 h minimum interval between injections for process machinery is a useful design figure to know.

As for the risk of over-purging the grease from a bearing, when the disc is mounted below it, Mr Hjertzen mentions this point also; the disc is not so effective that it pulls all the grease from the bearing. Certainly we saw no evidence of this in our tests. Further, it is no longer necessary for a bearing to contain grease in bulk for it to be effectively lubricated. As I said in my paper, all that is required from a modern grease is a film over the working parts—indeed, from this aspect, it is the intended function of a disc to pull all the bulk grease out of the bearing leaving it filmed only. In this way the bearing runs at its lowest temperature and consumes its lowest power. If churning takes place, temperature and power consumption rise. According to the degree of churning, the power consumed can increase two to ten times. This may not matter much for an odd bearing or so but if it happens in the thousands of bearings in a large factory, one can be making a generous contribution to the £500 million mentioned by Mr Jost.

I have no real experience of valve applications in the field and so I am unable to quote any practical details of value.

Dr Summers-Smith questions my re-lubrication interval of 300 h. I must emphasize that this figure relates to the pairs of bearings running at their recommended maximum speeds. At lower speeds, the interval can be safely increased. For example, if the bearings are confined to 80 per cent of their maxima, the interval then becomes the desired 500 h. I did say I was 'playing-safe' at half lubrication life. Bearings will not fail as a result of the over-use of a valve system but they may well fail from under-use. Why take the risk?

Mention is made of the possible switch to oil-lubrication systems. Such systems are likely to want more attention to keep them in order than a grease-valve system, apart from many other adverse features, e.g. cost, complication, drippage, and increased power consumption.

In am in accord with Dr Summers-Smith's desire to possess a standard form of reliably predicting grease-lubrication intervals. The variables involved are numerous and the task of formulating a procedure formidable. I have seen quoted somewhere that the accuracy of such predictions is likely to be ±50 per cent and I can well believe this.

I thank Mr Vose for his information, given in his first item, of which I was unaware. I have written above on the subject of dirty environments, his second item. A good thick wall of static grease forms an effective barrier against the ingress of dirt. I appreciate his point about a bend in the eject passage. I did say in my paper that the passage, if at all long, should be of increasing sectional area, i.e. flared. I agree that injections are equally effective whether the machine is running or standing. The point he makes, that grease inlet and ejection points should be positioned to suit the prevailing temperatures, is important, probably more so, in my view, at the higher temperatures than at the lower outdoor temperatures. Injection passages are, of course, full of grease, and if the machine is handling a hot process, the grease is being cooked. The next injection thus pushes cooked grease into the bearing, which is not a desirable event.

Mr F. E. H. Spicer—I should like to thank Mr Beardsley and Mr Clarke for their discussion of Paper 4.

Mr Beardsley's views were put before the IP Panel responsible for developing the test method. The members had accepted at the beginning of the work that they would not be able to produce a test method to satisfy all users and their aim was to produce a test which would fulfil the needs of as many as possible. The test was therefore designed to assess greases used under conditions where there was free water present but not where the water was frequently being changed. Typical conditions were those of bearings contained in poorly sealed housings subject to normal variations in atmospheric temperature and humidity. The operating conditions for British Railways rolling stock seem to be rather more severe than the test was intended to simulate since they appear to involve an element of water washing. Nevertheless I am sure that Mr Beardsley will know that the Emcor test, from which the IP 220 method was developed, was the anti-corrosion test recommended for use by the International Union of Railways in 1964.

The term 'not suspect' as used in the paper does not relate to the performance of the grease but to the way in which the test has been carried out. It indicates that the results themselves are acceptable, not that the performance level of the grease is suitable. Indeed, as was emphasized in the paper, the Institute of Petroleum does not consider setting performance levels to be part of its function. Its purpose is to produce well defined and unambiguous test methods which can be employed by users and producers of petroleum products in specifications which will ensure the uniformity and suitability of those products for any given application. This virtually precludes the use of the 'go–no go' type of test in IP standards, since such a test must assume a level of performance which is unlikely to be acceptable to all users of the method.

To say that the British Railways greases 1 and 2 are equivalent to greases A and C purely because the soap bases of the greases are similar is perhaps making a rather tenuous comparison. It is surely unwise to compare the IP test with the British Railways test on so little evidence. It would be interesting to assess the precision of the B.R. test by a programme similar to that used for the IP method but it is doubtful whether the time and effort could be justified for a method which has been designed for and correlated with this specific application.

In the IP 220/67 method the actual load, due to the weight of the shaft and inner race assemblies, is about 1 lb per bearing and this is distributed over nine balls. Each ball makes approximately 80 000 passes through the loaded zone during the first three days of the test. While we are not able to measure the grease-film thicknesses obtained by this method, it is believed that they are not too unrealistic.

The four members of the Panel from the bearing-manufacturing companies felt that, although some years ago some bearing steels may have been sensitive to the type of inhibitor used, this is not true of current materials. The Panel believe that the IP 220 method is a reliable anti-corrosion test with a wide field of application. There is no reason why the test should not be modified to make it more applicable to specific operating conditions. This could be done, for example, by replacing the distilled water with some more appropriate fluid, using test bearings from different manufacturing sources, or by changing the load, duration, or running conditions. It must be remembered, however, that any deviation from the method would mean that the quoted precision was no longer applicable. It would therefore become necessary for the user to establish this himself if he is to have a clear idea of the validity of the test.

With reference to Mr Clarke's discussion, there is always a danger that increasing the severity of a test may make it unrealistic. Tests with sodium chloride solution are obviously appropriate to such applications as automotive wheel bearings which may have become contaminated with salt from roads. There are several references in the literature to sodium chloride solutions producing corrosion of a different degree of severity and of a different type from sea water so that tests with sodium chloride solution will not necessarily simulate the effect of sea water contamination. Basically, I think that it would be wrong to increase the severity of the IP 220 test by using sodium chloride solution in place of the distilled water since this would inevitably narrow its field of application. However, I am completely in favour of using the appropriate contaminant, sodium chloride, sea water, mine water, etc., in the IP method to assess the rust protection offered by greases under specific operating conditions. It would then of course be necessary to establish the precision of the method under these conditions.

Ir G. J. Scholten—Generally speaking, Mr Dodson is right. If the shear stresses in the lubricating film are smaller than the yield strength, the shear zone will tend to become localized somewhere. This situation is possible when the effective viscosity is very low, for example as low as the base-oil viscosity, the shear rate is low, and the yield strength is very high. In that case the measured shear stresses have to be constant over a certain shear-rate region.

As can be seen from Figs 5.4 and 5.5 this can and will never occur, because the shear stress is always larger than the yield strength.

I hope this will answer Mr Dodson's question.

Mr Hutton is correct when he states that the cone-and-plate viscometer can leak considerably, especially at higher shear rates.

The cylinder type of viscometer used in our laboratory was provided with helix seals at both ends of the operating part to prevent leakage (see the General Notes on p. 36). Different tests on bearings provided with helix seals and using oil or grease as the lubricant, have proved that there is no detectable leakage after a few hours of operation.

As to the 'mirror test', such a test can be used indeed as a rough method of examination if a grease has a low yield strength. It is difficult to say, however, whether a mirror-like smoothness and flatness of a grease surface indicates that there is no yield strength at all, because bleeding can play an important part in the last stage of the flattening of the surface.

Mr J. G. M. Sallis and **Mr W. H. Wilson**—We accept Mr Harris's criticism that one cannot classify a grease fully by simply referring to the gelling agent. In defence we would point out that this is how the majority of grease users recognize what the oil companies sell them and also by looking at the trade name and reference number. For most practical purposes such an identification is quite adequate.

At the time of writing we decided not to include manufacturers' names and brand types for the greases investigated but this information is available from us on request.

Mr Shorten's reference to the addition of carbon fibres to grease as a solid lubricating phase is interesting, although we do not think the practice will become popular. There is no evidence that carbon in fibre form behaves in any way like flake carbon. Its value as a dry lubricant seems to be in the abrasive polishing of surfaces it produces.

We thank Mr Shorten for the information that his company is manufacturing polyethylene-treated greases to the U.S. prescription. We believe it is true, though, that attempts by other U.K. manufacturers to develop their own polyethylene-containing greases have been only moderately successful and the resulting lubricants have not shown the advantages claimed for the U.S. versions.

Mr A. E. Yousif and **Dr K. D. Bogie**—In reply to Mr Hutton, as yet we have made no attempt to apply non-linear models of viscoelasticity to the present work, but we hope to do this in the very near future, using evidence acquired from previous experimental work as a guide.

The second part of the question requires further explanation because of poor definition by us. Breakdown should, we believe, be restricted to soap-thickened greases where the secondary phase is relatively weak and consequently the thickeners break up into fibrils. In the present context breakdown refers to the disintegration of flocculated groups of particles, and not to the particles themselves.

Alignment and breakdown of the flocculated groups were not studied in situ, e.g. on the Weissenberg Rheogoniometer, but in a small shearing machine specially designed and adapted to a Vickers Polaroid microscope. The bottom stage of the shearing machine could be rotated at constant speeds of 0·1, 1·0, and 10 rev/min. An electronic circuit was designed such that the d.c. current from the motor could be fed to the illumination circuit on the microscope to provide stroboscopic pulses corresponding to the input speed of the bottom platen and so permit analysis of the material under shear.

The system briefly described was used to analyse the flow patterns of the various suspensions. These observations were used to interpret the results from the Weissenberg Rheogoniometer.

It is accepted that the two systems do not match and undoubtedly differences in flow condition will be experienced, but at the time no other way was open which would permit visual inspection of the sheared material. Isotope tracers were considered but rejected because of the complex procedures involved.

Because the resolution of the microscope system was limited to 1·5 μm, only groups of particles above this size could be studied. Nevertheless, visual and photographic evidence was obtained which illustrated the break-up of flocculated groups into smaller groups. It is reasonable to assume that submicron groups also break up as the imposition of shear is increased until a state is reached where break-up and reformation of the structure are constant.

We are of the opinion that the dynamic response of the synthetic grease described in this paper can be interpreted in a manner similar to that which Kofenbach [1] used to explain the inter-fibre forces (ionic and Van der Waal) and three-dimensional structure existing in soap-thickened greases.

If the synthetic-thickener particles are considered on an idealized basis, which means they are considered to be spherical (and they are very nearly, according to electron micrographs), of equal size, and non-porous, then the forces acting between the particles can, we suggest, be considered on an electrostatic repulsion basis.

If this view is accepted, then evidently the actual forces required to initiate movement must be greater than the inertia forces of the mass to be moved, the net Coulomb forces and the frictional forces which may or may not be present in the grease.

The maximum repelling force of each particle can be

considered as a function of the minimum inter-particle distance, and will depend largely upon the magnitude of the repelling forces between the particles at the minimum distance. The inter-particle distance will, we suggest, depend upon two factors, the concentration of the particles and the size of the particles. Consequently the magnitude of the repelling force at a given inter-particle distance will be a combination of particle-surface state and the electrolytic properties of the base fluid.

If this theory is applied to the type of curves obtained for the synthetic grease investigated here, it is evident that the forces required to initiate movement of the structure will be dependent upon the sum of the maximum repulsive forces between the particles being moved. Under such a condition the acceleration and velocity of the mass of grease being moved is very low, and therefore the force required to initiate movement will depend directly upon the concentration of particles, size, and surface state, together with the electrical properties of the fluid.

Once flow has been initiated, the total force required to continue the movement of the mass of grease will be reduced, because in effect the momentum acquired by the particles will carry them beyond a position of minimum repulsion. It is evident, therefore, that the force required to maintain a constant rate of shear will be reduced by an amount equal to the inertia and momentum forces of the system.

REFERENCE
(1) FOSTER KOFENBACH *NLGI Spokesm.* 1956 **20** (No. 3), 16.

Dr E. D. Yardley and **Mr W. J. J. Crump**—We appreciate the validity of Mr Langston's remarks, and it has always been our aim to visit the 'scene of the crime' whenever possible. However, we would add that, while visits to the field by laboratory specialists are always desirable, they are not always possible. In such cases examination of failed components in the laboratory is surely preferable to no examination at all.

Mr H. D. Moore, Mr J. W. Pearson, and **Mr N. A. Scarlett**—We believe that the sales resistance experienced by Mr Shorten to greases which do not fit the NLGI classification follows only because grease marketers have for many years adhered to the NLGI system. Thus their customers have become used to the NLGI grades. Unfortunately, because of the absence of overlap between grades, there is no opportunity in the NLGI classification for 'multi-grade' greases corresponding to the 'multi-grade' oils fitting the SAE crankcase oil classification. However, it is our experience that a grease outside the NLGI classification can be marketed successfully if it is shown that the intermediate penetration is technically desirable. The evidence needs to be from actual service experience and not just from laboratory or rig tests.

As for resistance to approvals, much depends on the approving authority. If the specification is for an NLGI grade, it is not surprising if the approving authority refuses off-grade products. There are, however, many specifications which are not bounded by NLGI grades, e.g. CS 3107B (penetration range 250–300) and DGS 6921A (penetration range 200–300).

On the last point we have not found any simple correlations between the properties mentioned and rig (or service) performance. The viscosity at a high rate of shear is much dependent on the oil viscosity; it also depends on soap content: the yield stress is also much dependent on soap content but varies with temperature in a way dependent also on soap type and method of manufacture. Even so, grease structure can be over-riding in particular rigs or bearings.

In reply to Mr Vose we would say that, as 'boffins' once removed, we are always grateful for opportunities to build up our stock in the eyes of our maintenance-engineering colleagues. Unfortunately, as we pointed out in the early part of our paper, our early efforts with the motors in question were disastrous. While this led us to forsake simple laboratory tests and turn to more realistic evaluation in engineering components, we must consider ourselves fortunate that our refinery maintenance colleagues were sufficiently understanding to allow us a final chance to prove the virtues of grease when properly developed. More seriously, we believe that the problems in these particular motors provided a salutary experience for bearing manufacturer, motor designer, and grease manufacturer. This served to underline the necessity for close co-operation between the three parties at an early stage in the development of advanced components. Without this close co-operation, the machine operator, who is the customer of all of us, suffers—and so do the reputations of his suppliers!

Mr Hutton points to a link joining our practical paper with the more fundamental papers Nos 1 and 8. We believe many more links will be needed between fundamental and practical work before we can achieve our objectives of fully relating grease performance in bearings to grease structure.

As Mr Harris's contribution covers the whole Symposium, and not just our paper, we may perhaps be permitted to extend our reply. We agree that the affect of grease structure on grease performance is not widely recognized. We believe that recognition of this factor would explain some of the differences in performance between various greases which many investigators have found inexplicable in terms of normal physical and chemical properties. The problem of quantifying the structural factor leads us to believe that Mr Harris's proposal for a B.S.I. code classification of greases for communication purposes would be of little value.

On compatibility, we agree that the over-heating of a bearing when relubricated may well, in some circumstances, have been attributed to 'mixing of the greases', i.e. incompatibility. However, we suggest that during the hot-running period, which is caused by redistribution of the excess grease, sufficient physical mixing and working of the two greases may occur on a wider scale than Mr

Harris postulates and thus lead to the same changes in significant properties as are observed after deliberate mechanical mixing of greases in the laboratory. Thus, where the two greases can be shown to be incompatible when mixed, we believe there will be an element of risk in mixing them in a bearing, and this should be avoided for difficult applications. There is of course less possibility of trouble where the conditions are not difficult, or where re-lubrication is done frequently as with a centralized system.

Summary of Discussion on Report Papers

THERE WERE lively discussions after the presentation of the Report papers, and the following are some of the comments that were made. Mr R. A. Clarke of Luton, Beds., questioned Mr R. Cosher on his statement that grease component bearings had to have an optimum life of at least four years of aircraft service. This did not take account of the fact that many bearings were held in stores for five to seven years, and the bearing supplier could not be held responsible for grease deterioration over such a long period. B.O.A.C. were well aware of this problem and in such circumstances the bearings were cleaned and re-greased before assembly in the aircraft.

Mr Clarke also asked Mr I. S. Roberts whether any detailed work had been undertaken to evaluate the lubricating properties of the soap-thickened diester fluids currently being used as lubricants for 'greased-for-life' gear units. Mr Roberts replied that no bearing problems had been encountered during the gear tests and it was therefore assumed that the lubricant was quite satisfactory for the bearings.

Two points were raised after the report by Mr Hjertzen and Mr Clarke, the first being whether the authors could give further details on the use of the minimum lubrication systems for machine-tool spindle bearings. They replied that experiments had been carried out to establish the minimum temperature rise which could be experienced, particularly in jig borer spindles. Very small quantities of grease, for example 1 g for every 50 mm bore of a cylindrical roller bearing, were added by means of a syringe and smeared carefully around the bearing surfaces. In this manner it had been possible to run a 90-mm-bore roller bearing in a jig borer operating at 2000 rev/min with a maximum temperature rise of 8 degC. Quite obviously, such a technique relied on planned maintenance because it was not possible to add grease during the normal service life.

The second point raised was when bearing manufacturers were going to standardize their recommendations for grease re-lubrication periods. Mr Hjertzen pointed out that earlier in the Symposium it had been shown that there was a very close relation between the re-lubrication periods given by the bearing manufacturers and grease suppliers. Moreover it must be realized that when bearing manufacturers put forward their recommendations they must be applicable to all greases and not only the premium types. Consequently, in the ultimate the answer did not lie with the bearing manufacturers but rather in the standardization of greases and grease test rigs. It was, however, quite obvious that as research developed the re-lubricating recommendations would become more standardized.

Mr R. P. Langston, of Cobham, Surrey, asked if Dr Marini, author of the short paper on 'The lubrication of office equipment', could say if his studies indicated that only a limited number of lubricants were needed to lubricate office machinery (perhaps only two—one oil and one grease) or whether he was finding it necessary to specify a multitude of different products for different parts of the equipment. Dr Marini replied that his present studies indicated that a very limited number of lubricants were needed to lubricate office equipment. He was trying to reduce the lubricants to two: an oil and a grease. He had developed a unique oil which was extensively used on production lines, and hoped he would have a unique grease by the end of the year (1970). Obviously, in special cases other lubricants would be necessary, but only in very small quantities.

List of Delegates

BANFORD, W. J.	James H. Heal & Co. Ltd, Halifax, Yorkshire.
BARNETT, R. S.	Texaco Inc., Beacon, N.Y.
BEARDSLEY, J.	British Railways, Derby.
BLOUNT, G. N.	Michelin Tyre Co. Ltd, Stoke-on-Trent.
BRYANT, H. F.	Tecalemit (Engineering) Ltd, Plymouth, Devon.
CARLSSON, K. R.	SKF, Gothenberg.
CLARKE, R. A.	The Skefco Ball Bearing Co. Ltd, Luton, Beds.
CRAWLEY, P.	Castrol Research Laboratories, Bracknell, Berks.
CRUMP, W. J. J.	G.E.C. Power Engineering Ltd, Whetstone, Leicester.
DAWES, M. I.	*Tribology*, Guildford, Surrey.
DAWREY, S.	Shell-Mex and B.P. Ltd, London.
DAY, C. E.	National Coal Board, Doncaster, Yorkshire.
DAY, P. C.	Tecalemit (Engineering) Ltd, Plymouth, Devon.
DODSON, S. C.	B.P. Research Centre, Sunbury-on-Thames, Middx.
DRURY, N.	Appleby-Frodingham Works, British Steel Corporation, Scunthorpe, Lincs.
DYSON, A.	Shell Research Ltd, Chester.
EDWARDS, A. A. W.	Fighting Vehicles Research and Development Establishment, Chertsey, Surrey.
ERICSON, G.	Atlas Copco AB, Stockholm.
FARID, M.	Michell Bearings Ltd, Newcastle upon Tyne.
FARNFIELD, N. E.	Cranfield Institute of Technology, Cranfield, Bedford.
FARRELL, R. P. C.	Beaconsfield, Bucks.
FASCIANO, D.	Olivetti & C., S.p.A., Ivrea, Italy.
FLANDERS, B. M	Swansea, Glam.
FOOT, C. W.	Chelmsford, Essex.
FOWLE, T. I.	Shell International Petroleum Co. Ltd, London.
FRENCH, E. L.	Taylor Woodrow Construction, Southall, Middx.
GEACH, C. J.	B.P. Research Centre, Sunbury-on-Thames, Middx.
GROSZEK, A. J.	B.P. Research Centre, Sunbury-on-Thames, Middx.
HARRIS, J. H.	Chichester, Sussex
HARVEY, J. H. D.	Midland Silicones Ltd, Barry, Glam.
HJERTZEN, D. G.	The Skefco Ball Bearing Co. Ltd, Luton, Beds.
HOBBS, R. A.	Ransome Hoffmann Pollard Ltd, Newark, Notts.
HOBSON, W. C.	British Steel Corporation, Scunthorpe, Lincs.
HUDSON, J. B.	Shell-Mex and B.P. Ltd, London.
HUMPHREYS, A. V.	Skefko Ball Bearing Co. Ltd, Luton, Beds.
HUTTON, J. F.	Shell Research Ltd, Chester.
JACKSON, A. G.	Farvalube Ltd, Hereford.
JAGGER, E. T.	George Angus & Co. Ltd, Wallsend, Northumberland.
JEMMETT, A. E.	Paisley College of Technology, Paisley, Renfrewshire.
JONES, E. F.	Esso Petroleum Co. Ltd, Abingdon, Berks.
JONES, M. F.	Imperial Chemical Industries Ltd, Billingham, Teesside.
JONES, R. H.	Basildon, Essex.
DE JONGE, H. F.	Shell Ned. Verkoopmij N.V. MIT, Rotterdam.
JORDAN, A. D.	Iliffe Science & Technology Publications, Guildford, Surrey.
KEMP, R. A.	University of Leicester.
KENMIR, G.	Darlington, Co. Durham.
KEYSELL, M.	Liverpool Regional College of Technology.
KINNER, G. H.	Ministry of Technology, London.
LANGBORNE, P. L.	National Engineering Laboratory, East Kilbride, Glasgow.
LANGSTON, R. P.	Admiralty Oil Laboratory, Cobham, Surrey.
LANSDOWN, A. R.	Swansea Tribology Centre.
LEWIS, N. J.	Shell-Mex and B.P. Ltd, Leeds.
LIDÉN, L. O.	Gulf Research Lab. N.V., Rotterdam.
MACKENZIE, K.	Imperial Chemical Industries Ltd, Stevenston, Ayrshire.
MALMBORG, E. T.	London.
MARINI, M.	Olivetti & C., S.p.A., Ivrea, Italy.
MERRETT, J. G.	Centralube Ltd, London.
MIDDLEMISS, S. L.	Mobil Oil Co. Ltd, Stanford-le-Hope, Essex.
MILTON, R. J.	The Skefco Ball Bearing Co. Ltd, Luton, Beds.
MOORE, H. D.	Shell Research Ltd, Chester.
MORGAN, A. W.	Admiralty Oil Laboratory, Cobham, Surrey.
MORRIS, N. R. W.	Farvalube Ltd, Hereford.
MORTON, I. S.	University of Birmingham.
MULLETT, G. W.	Ransome Hoffmann Pollard Ltd, Newark, Notts.
McALPINE, A. F.	J. H. Fenner & Co. Ltd, Hull, Yorkshire.
NAPIER, G.	Shell-Mex and B.P. Ltd, London.
NOWAK, J. Z.	K. S. Paul Products Ltd, London.
O'DONNELL, P. J.	Elliott Brothers (London) Ltd, Camberley, Surrey.
PALMER-LEWIS, I.	Shell-Mex and B.P. Ltd, London.
PATERSON, E. V.	*Industrial Lubrication and Tribology*, Brosley, Shropshire.
PEARSON, J. W.	Shell International Petroleum Co. Ltd, London.
POMLETT, M.	Shell-Mex and B.P. Ltd, Birmingham.
POON, S. Y.	University of Cambridge.
PORRVIK, M. E.	AB Svenska Shell, Stockholm.
RANDLE, J. N.	Central Research Laboratory, Coventry, Warwickshire.
ROBERTSON, J. A.	Shell-Mex and B.P. Ltd, London.
ROBINSON, N.	Worcester Park, Surrey.
SCARLETT, N. A.	Shell Research Ltd, Chester.
SCOTT, D.	National Engineering Laboratory, East Kilbride, Glasgow.
SCOTT, G. W.	London.
SHORTEN, G. A.	Walkers (Century Oils) Ltd, Hanley, Stoke-on-Trent.
SLAUGHTER, T. M.	The Post Office, London.
SMITH G.	Luton, Beds.
SMITH, G.	Shell B.P. Scotland Ltd, Glasgow.
SOUL, D. M.	Lubrizol International Laboratories, Derby.
SPICER, F. E. H.	B.P. Research Centre, Sunbury-on Thames, Middx.
STOCK, B.	Railko Ltd, High Wycombe, Bucks.
SUDDABY, O.	Acheson Colloids Company, Plymouth, Devon.
SUMMERS-SMITH, D.	Imperial Chemical Industries Ltd, Billingham, Teesside.
SYMMONS, E. A. M.	Mobil Oil Co. Ltd, Stanford-le-Hope, Essex.
TAYLOR, J. D.	V.S.G. Engineering Ltd, Slough, Bucks.
TRITTON, T. L. J.	Texaco Ltd, London.
VAESSEN, G. H. G.	Metaalinstituut TNO, Delft, Holland.
VAMOS, E.	Institut NAKI, Budapest.
VAN DEN BERG, E. G.	B.I.P.M., Holland.
VAN DOORNE, G. C. L	Labofina, S.A., Brussels.
WILSON, R. A.	Shell Research Ltd, Chester.
WILSON, W. H.	The University of Leeds.
WYLLIE, D.	Admiralty Oil Laboratory, Cobham, Surrey.
YARDLEY, E. D.	National Coal Board, Burton-on-Trent, Staffs.
YOUSIF, A. E.	University of Salford.

Index to Authors and Participants

The names of authors and the numbers of pages on which papers begin are printed in bold type.

Barnett, R. S. 87
Beardsley, J. 94
Bogie, K. D. 57, 104

Clarke, R. A. vii, 94, 107
Cosher, R. vii, 107
Crump, W. J. J. 63, 105

Day, P. C. vii
Dodson, S. C. 94
Dyson, A. 1

Foot, C. W. vii
Fowle, T. I. vii

Harris, J. H. 94
Hjertzen, D. G. vii, 95, 107
Hutton, J. F. 12, 96

Jackson, A. G. vii

Langborne, P. L. 82
Langston, R. P. vii, 97, 107

MacKenzie, K. vii
Marini, M. vii, 107
Moore, H. D. 74, 105
Morgan, A. W. 48
Morris, N. R. W. 97
Mullett, G. W. 17, 102

Neale, M. J. vii

Parkinson, D. R. vii
Pearson, J. W. 74, 105
Pike, W. C. vii
Poon, S. Y. 97

Roberts, I. S. vii, 107
Robertson, J. A. vii

Sallis, J. G. M. 40, 104
Scarlett, N. A. 74, 105
Scholten, G. J. 32, 103
Scott, D. vii
Shorten, G. 99
Spicer, F. E. H. 23, 103
Summers-Smith, D. 100

Trevalion, P. O. vii

Vose, J. K. vii, 100

Wilson, A. R. 1
Wilson, W. H. 40, 104
Wyllie, D. 48

Yardley, E. D. 63, 105
Yousif, A. E. 57, 104

Subject Index

Titles of papers are in capital letters.

Additives; 88
 review of American literature on, 104
Agricultural machinery, grease lubrication, review of recent American papers on, 88
Aircraft; bearings, service life of, 107
 review of recent American papers on grease lubrication, 88
Aluminium alloys, as bearing materials, comparison of performance with oil and with grease lubrication on Beta test rigs, 45, 95, 99, 104
America, United States of, review of some recent publications on lubricating grease, 87
APPARENT (DYNAMIC) VISCOSITY AND YIELD STRENGTH OF GREASES AFTER PROLONGED SHEARING AT HIGH SHEAR RATES, 32
Asbestos, value of its addition to grease, 46
ASLE Transactions, review of recent American lubricating grease literature published in, 87
Automatic equipment lubrication, review of recent American papers on, 88
Axle boxes, railway wagons, review of recent British papers on overheating, 84

Bearings, lubrication, *see all entries*
Beta test machine and performance tests on bearing material/grease combinations, 42, 95, 99, 104
Brakes, review of recent American papers on lubrication, 88
Brinelling, as cause of rolling bearing failure, 50, 53
British Railways; 'go-no go' corrosion tests on grease lubrication for outdoor equipment: 94
 comparison with results from IP 220/67 tests, 94, 103
 problem of hot axle boxes in wagons, review of recent British papers on, 84
Bulk handling systems, grease lubrication, review of recent British papers on design, 84

CALCULATION OF THE EFFECT OF THE COMPRESSIBILITY OF GREASE ON THE PERFORMANCE OF A TWIN-LINE DISPENSING SYSTEM, 12
Carbon blacks, tests on properties as thickening agents in synthetic greases, 57, 96, 104
Carbon fibres, value of their addition to grease as a solid lubricating phase, 99, 104
Centralized grease systems, review of recent British papers on design, 84
Classification of greases; desirability of, 95, 104, 105
 review of recent American papers on, 88
'Clearability' of greases, relation to structure, 76, 96, 105
Coal mine lubrication, review of recent American papers on, 88
Communications industry, review of recent American papers on, 88
Compatibility of greases; 1, 95–98
 incompatibility as probable cause of bearing failures, 73, 95
Compressibility of grease; calculation of effect on performance of a twin-line dispensing system, 12
 in pipelines, determination of, review of recent British papers, 83
Cone Resistance Value as a measure of yield stress in grease, 77
Contamination, as cause of bearing failure, 48, 55, 65, 97, 99, 100, 102
Co-ordinating Research Council Method L-41-1957 for assessing anti-rust properties of lubricating greases, 24, 94, 103
Copper, for bearings, comparison of performance with oil and with grease lubrication on Beta test rig, 45, 95, 99, 104
Corrosion; British Railways 'go-no go' tests on grease lubrication for outdoor equipment, 94
 effect on bearings, and IP test 220/67 for assessment of rust-prevention characteristics of greases, 23, 94, 103
 review of recent American papers on, 88
 survey of bearing failures caused by, ships' electrical machinery, 50, 95, 97, 99, 101
Cost, ratio, various greases, 46
Couette viscometer, design and procedure for measurement of apparent viscosity of grease, 32, 94, 96, 103

Deposit formation, elimination in development of high-performance grease for industrial rolling bearings, 78, 95, 96, 99, 101, 105
Design; centralized, and bulk handling, grease lubrication systems, review of recent British papers on, 84
 Libra, Beta, and Zircon testing machines for bearing material/grease combinations, 42
 rolling bearings 'Standard Life' (ISO R/281), 100
 twin-line lubricant dispensing system, calculation of effect of compressibility of grease, 12
 two-bearing grease valve mountings, with efficiency test results, 17, 95, 102
 viscometer and yield strength meter, 32, 94, 96, 103
DESIGNS FOR MOUNTINGS INCORPORATING TWO ROLLING BEARINGS AND A GREASE RELIEF SYSTEM, 17
Development problems, high-performance grease for industrial rolling bearings, 74, 95, 96, 99, 101, 105
Disc machine measurement of film thicknesses in elastohydrodynamic lubrication of rollers by greases, 1, 94–98
Discs, valve, positioning in grease relief systems, 17, 95, 102
Dispensing systems; bulk handling and centralized, review of recent British and American papers on, 84, 88
 twin-line, calculation of effect of compressibility of grease for design, 12

Electric motors; refinery, development and performance of grease for, 81, 101, 105
 review of a recent British paper on grease tests for, 84
Electrical machinery, ships', survey of failures of grease lubricated rolling bearings; 48, 95, 97, 99, 101
 methods of examination, the failure pattern and most frequent causes, 48, 49, 51, 52, 55
Emcor anti-rust test, lubricating greases, as basis for IP method 220/67, 24, 94, 103
European participation in international precision evaluation programme of anti-rust properties of grease, and results, 25, 94, 103

Failures, grease-lubricated bearings; rolling, ships' electrical machinery, survey of investigations covering mainly corrosion, machine and fitting defects, and contamination, 48, 95, 97, 99, 101
 renewal of lubrication by the grease valve method, with designs of double-bearing mountings, 17, 95, 100, 102
 rolling-element, investigation of failures for which grease or grease and other factors were responsible, 63, 95, 97, 105
FAILURES OF GREASE-LUBRICATED ROLLING-ELEMENT BEARINGS, 63
Fatigue, as cause of bearing failure, 50, 64
Ferranti–Shirley cone-and-plate viscometer measurement of apparent viscosity of grease, 96, 104
FILM THICKNESSES IN ELASTOHYDRODYNAMIC LUBRICATION OF ROLLERS BY GREASES, 1
Fitting defects causing failure of rolling bearings, 52
Flow of grease; calculation of in pipes for twin-line dispensing system design, 12
 review of recent American papers on, 88
Food industry, American research on lubricating grease, 88

SUBJECT INDEX

Fuel pump bearing failures due to incompatibility of fluid and bearing grease, 65

Gear lubrication, review of recent American papers on, 88
Glacier DX plastic bearings, tests on grease lubrication and effect of various greases on performance; 42, 43, 95, 99, 104
    comparison with common metal bearing materials tested on Libra machine, 43
GREASE LUBRICATION: A REVIEW OF RECENT BRITISH PAPERS, 82
GREASES AS LUBRICANTS FOR METAL AND PLASTIC-LINED PLAIN BEARINGS, 40

History, grease development, review of recent American papers on, 88
Hoffman rig tests, development of grease for industrial rolling bearings, 75

IP DYNAMIC ANTI-RUST TEST, LUBRICATING GREASES, 23
Iron and steelworks lubrication, review of recent papers on; American, 88
    Third Annual Meeting of the Lubrication and Wear Group, 82
ISO R/281 'Standard Life' basis for rolling bearing design, 100

Klein mill tests on effect of mechanical degradation on grease film thickness, 97

Libra test machine for assessment of bearing material/grease combinations, 42, 95, 99, 104.
Load, excessive, as cause of bearing failure, 50, 53
Lubrication, *see all entries*
Lubrication Engineering, review of recent American lubricating grease literature published in, 87

Machine and fitting defects, causing failure of rolling bearings, 52
Machine-tool spindle bearings, minimum lubrication systems, 107
Manufacturing and processing of greases, review of recent American literature on, 88
Marketing of grease, review of recent American papers on, 88
Mechanical stability, synthetic grease, improvement of, 61
Metals, bearing, lubrication with oil and with grease, comparison of performance on Beta test rig, 45, 95, 99, 104
Military equipment, review of recent American papers on grease lubrication, 88
Mirror test, evaluation of yield strength of grease, 96, 104
Misalignment of bearings causing failure, 50, 70
Motor vehicles, review of recent American literature on grease lubrication, 88
Mountings incorporating two rolling bearings and a grease relief system; designs for, 17, 95, 100, 102
    operation of grease valves and their advantages, 18

National Lubricating Grease Institute; classification of greases, 99, 105
    summary of recent lubricating grease literature published in the *Spokesman*, 87
Nuclear power plant lubrication, review of recent American papers on, 88

Office equipment lubrication, 107
Oil and grease, comparison of film thicknesses formed by each in disc machine, 1, 94–98
Oil versus grease, tests on bearing material performance, 44

Packaging of greases, review of recent American papers on, 88
Paper mill lubrication, review of recent American papers on, 88
Particle size reduction methods, sodium nitrite added to lubricating grease; 80
    effect of particle size on grease performance, 80
Petroleum, Institute of; rig test IP 168, use in development of grease for industrial rolling bearings, 75
    test method IP 220/67, assessment of rust-prevention characteristics of greases: 23, 94, 103
    possible use in tests on failed bearings from ships' electrical machinery, 99
Photomicrographs of failed bearings, 72
Pipes, compression of grease in; effect on twin-line dispensing systems, 12
    review of recent British papers on, 83

Plastic bearings, grease lubrication, and performance tests on Libra machine; 42, 95, 99, 104
    comparison with performance of common bearing materials, 43
Polyethylene-treated greases, performance in plain bearings, 46, 99, 104
Polytetrafluorethylene-filled greases, performance in plain bearings, 46, 99, 104
PROBLEMS IN THE DEVELOPMENT OF A HIGH-PERFORMANCE GREASE FOR INDUSTRIAL ROLLING BEARINGS, 74
Properties of grease, review of; papers at Lubrication and Wear Meeting on iron and steelworks lubrication, 82
    publications on American research, 88
Pump, fuel, bearing failures caused by incompatibility of fluid being pumped and lubricating grease, 65

Railways; *see also* British Railways
    review of recent American papers on grease lubrication, 88
Ransome and Marles, rig tests, development of grease for industrial rolling bearings, 75
Rationalization of lubricants, review of recent British papers on, 85
Refineries; development of grease for electric motors and performance of, 81, 101, 105
    review of recent American papers on, 88
Review of recent British and American grease lubrication papers, 82, 87
RHEOLOGICAL BEHAVIOUR OF A NEW HIGH-TEMPERATURE SYNTHETIC GREASE, 57
Rheology of greases, 57, 88, 94
Rollers, elastohydrodynamic lubrication, disc machine tests on film thickness, 1, 94–98
Royal Navy, *see* Ships
Running, rough, small bearings, development of grease to obviate, 74, 80
Rust prevention properties in grease; assessment of by IP 220/67 test, 23, 94, 103
    development of for industrial rolling bearings, 78, 95, 96, 99, 101, 105
    review of recent American papers on, 88

Salt solution, for simulation of sea-water contamination, undesirability of use in IP 220/67 anti-rust tests on grease, 24, 94, 103
Ships; Royal Navy, survey of failures of grease lubricated rolling bearings in electrical machinery, examination methods and causes, 48, 95, 97, 99, 101
    recent American papers on, review of, 88
    use of polyethylene glycol for reduction of skin friction of hulls, 46
Skefco rig tests; development of grease for industrial rolling bearings, 75
    dynamic anti-rust, 29
'Snowballing', designing to avoid in twin-line dispensing systems, 12
Sodium nitrite, incorporation into grease and beneficial effect on performance of industrial rolling bearings; 79, 95, 96, 99, 101, 105
    methods for reduction of particle size, 80
Space vehicles, review of recent American papers on grease lubrication, 88
Speed, operational, bearings; advantages of a grease valve, 17, 95, 102
'Standard Life' (ISO R/281) design basis, 100
'Standard Life' (ISO R/281) design basis for rolling bearings, 100
Standardization of lubricating recommendations; 100, 103, 107
    review of recent American papers on, 88
Stone industry, review of recent American papers on lubrication, 88
Structure of greases, 3, 4, 5, 43, 46, 48, 57, 74, 82, 87, 94, 97, 99, 103, 104
SURVEY OF ROLLING-BEARING FAILURES, 48
Synthetic grease, tests on rheological behaviour, 57, 96, 104

Tanks, amphibious, addition of asbestos to lubricating grease, 46
Temperature; bearings, operational: advantage of use of grease valves, 17, 95, 102
    design basis 'Standard Life' (ISO R/281), 100
    considerations in specification of ideal grease for plain bearings, 46, 95, 99, 104
    investigation of hot-running, high-speed bearings, 66, 76, 95, 96, 99, 101, 105

SUBJECT INDEX

hot axle boxes in railway wagons, review of recent British papers on, 84
stability of synthetic grease, improvement of, 61
Tests; British Railways 'go–no go' on corrosion of grease lubricated outdoor equipment, 94
 carbon blacks as thickening agents in synthetic greases, 57, 96, 104
 Co-ordinating Research Council Method L-41-1957 and Emcor, conditions, and choice of latter as basis for IP 220/67, 24, 94, 103
 development of high performance grease for industrial rolling bearings and rigs used, 74, 75, 95, 96, 99, 101, 105
 effect of grease type on performance of plastic bearings, 43
 efficiency of various designs of two-bearing grease valve mountings, 17, 95, 100, 102
 for cause of failure of rolling bearings, ships' electrical machinery, 48, 95, 97, 99, 101
 Institute of Petroleum Method 220/67, anti-rust assessment of greases, 23, 26, 94, 103
 investigation of overheated failed bearings, 68
 laboratory methods for grease, review of recent British papers on, 84
 Libra, Beta, Zircon, test machines for bearing material/grease combinations, 42, 95, 99, 104
 'mirror', evaluation of yield strength of grease, 96, 104
 review of recent American publications, 88
 rheological behaviour, high-temperature synthetic grease, 57, 96, 104
 viscosity and yield strength of grease, with Ferranti–Shirley cone-and-plate, and Couette type viscometers and yield strength meter, 32, 94, 96, 103, 104

Theory of the ideal grease for plain bearings, 40, 45, 95, 99, 104
Thickeners; American papers on, review, 88
 carbon blacks, tests on addition to synthetic greases, 57, 96, 104
Thornton rig tests, development of grease for industrial rolling bearings, 75
Thrower collars, see Discs

Valve type, grease relief system, designs for mountings incorporating two rolling bearings, 17, 95, 100, 102
Viscometer; Couette, design of, and test procedure for measurement of apparent viscosity of grease in once-lubricated bearings, 32, 94, 96, 103
 Ferranti–Shirley cone-and-plate viscosity investigations, 96, 104
Volkswagen 'Beetle' engine performance and its divergence from bearing metal/lubricant test results, 44

'Water-marking' of failed bearings failing from corrosion, 52
Wear, high-speed industrial bearings, development of grease to prevent, 80, 95, 96, 99, 101, 105
Weissenberg rheogoniometer, use in investigation of rheological behaviour of high-temperature synthetic grease, 58, 96, 104

XG-274 grease for Royal Navy rolling bearing lubrication, survey of failures, 48, 95, 97, 99, 101

Yield strength meter, design and tests on grease, 35, 94, 96, 103

Zircon test machine for ranking of bearing materials, 42, 95, 99, 104